D0722913

Contract Suretyship

Contract Suretyship

From Principle to Practice

Richard C. Lewis

John Wiley & Sons, Inc.

NEW YORK / CHICHESTER / WEINHEIM / BRISBANE / SINGAPORE / TOROI

Copyright © 2000 by John Wiley & Sons, Inc. All rights reserved.

Published simultaneously in Canada.

This publication is designed to provide accurate and authoritative information in regard to the subject matter covered. It is sold with the understanding that the publisher is not engaged in rendering professional services. If professional advice or other expert assistance is required, the services of a competent professional person should be sought.

Library of Congress Cataloging-in-Publication Data:

Lewis, Richard C.
    Contract suretyship : from principles to practice / Richard C. Lewis.—2nd ed.
        p.   cm.
    Includes index.
    ISBN 0-471-37135-1 (alk. paper)
    1. Insurance, Surety and fidelity—United States.   2. Contractors—Bonding—United States.   3. Construction contracts—United States.
I. Title.
K1228.L48   2000
343.73'078624—dc21   99-33864

Printed in the United States of America.

10 9 8 7 6 5 4 3 2 1

Contents

Acknowledgments

I would like to thank and acknowledge the following contributors who played a significant role in the creation of this book:

Teresa Davis Thamer, Certified Public Accountant, for her invaluable contribution in guiding me through many of the more technical construction accounting disciplines—as well as her editorial assistance in compiling Chapters 7 through 9.

My sister, Joan K. Lewis, for her expertise in proofreading, cutting, patching, and restructuring this material in order to provide greater continuity and clarity of the subjects covered.

Introduction

Without a mandated requirement, what construction firm would go eagerly shopping for a contract surety bond? Why would it pay for something it didn't need? The answer is, of course, it wouldn't. The demand for contract bonding usually arises from one of three sources: (1) federal, state, and local statutes involving public work; (2) the owner/developer and/or its construction lender on private work, where the lender may wished to be named in the bond as a second party at interest (dual obligee); or (3) a general or prime contractor requiring a bond of its subcontractors, at its own behest and/or that of its own surety company.

Unlike the manufacturer of widgets, who, under the laws of supply and demand, must compete in an open market place for promotion and the sale of its product, the surety industry already has a ready-made demand for its services. Conversely, its contractor clients must compete on what could hardly be called a level playing field compared to their typical manufacturing counterparts—both in terms of the disparate pricing paradox for their product, and the time allotted for its production. With the most basic cost control accounting system, the widget manufacturer knows almost to the penny the cost of its widgets and the overhead burden to be allocated before they are offered for sale. As the cost increases that increase is immediately passed along to the consumer. In the manufacturing organization there is no staffing designation for the position of an *estimator.*

A construction firm, on the other hand, won't know its final costs for months or years from the beginning of a project, during which time a multitude of unanticipated developments can drastically alter the original *estimate.* The elements of risk for the contractor and surety are, therefore, extraordinary and can be volatile for both, as documented loss experience for the surety industry most vividly attests.

For the benefit of those countless construction firms that are required to provide corporate surety bonding, the object of this book is to bring into sharper focus the underwriting psychology behind some of the surety's more questionable, confusing, and occasionally controversial decisions. It should, as examples, also provide greater depth and understanding of just how a construction firm can anticipate its surety's reaction to the diversion of its capital assets into nonoperational and extraneous enterprises, erratic and inconsistent profit/loss margins, as well as numerous other operating nuances affecting the levels of bonding capacity the company would be willing to extend.

Working with contractors in establishing, maintaining, and enhancing their surety bonding capacity for over 32 years has been an extremely rewarding personal experience, as well as a constant, ongoing learning process for myself and many of my contractor clients. Very little remains static in the construction field, particularly during times of economic downturns and scarcity of both public and private work. The occasional acrimony between the general contractor and his or her subs, suppliers, owners, and architects, and/or default by private owners in inadequately funding their projects, or the grim specter of unbonded defaulting subcontractors—together with a host of other incalculable complications—keeps the successful surety company ever alert to this myriad of unanticipated developments, and how they may ultimately affect its liability.

This book deals with a number of "real world" scenarios in the surety's selection process. Many of these case studies are based upon actual underwriting exercises experienced with existing and prospective contract accounts over many years, and have evolved into the nucleus of this text's principal theme and objective.

Beyond the financial aspects covered, other surety considerations are depicted in various case studies. Character and capacity, together with organizational depth, continuity plans, geographical job locations, political dynamics, digression from established building expertise into other construction fields, and involvement in speculative building ventures are just a few of the many other factors discussed.

It is these areas of conflict with traditional underwriting disciplines that often give rise to friction and confusion between a surety and its contractor. This text attempts to clarify the surety's rationale in taking what might well be a controversial position in conditionally authorizing or declining a bond request. Sadly, all too often, these positions are not articulated clearly and logically to the contractor by the surety representative, and the surety's response devoid of any effort to find a means of overcoming the objectionable features. The case studies illustrate how, with a positive approach and problem solving techniques, remedies may often be found in reaching mutually agreeable alternatives to whatever negative features may exist. In other cases, there will be no solution, and the reader will understand why there is not.

While this material is necessarily presented from the surety underwriter's perspective, it should give the reader a keener sense of how his or her bonding company may react to adverse financial and construction trends, and enable the contractor to initiate independently whatever remedial reforms may be required to rectify the problems(s) before the surety imposes its own solutions.

A final word of caution: As in any straightforward business association, where good faith dealings are paramount to the venture's success, no detrimental facts must be withheld. There should always be open and frank disclosures to the surety of any major problems the contractor is encountering—rather than having them appear in the next financial statement. It would be even worse for problems to be revealed by a barrage of claims from unpaid subs and suppliers or—even more devastating—by receiving these claims on top of default notices from owners.

Many of the widely accepted underwriting imperatives of 40 years ago have long since faded into obscurity. Some of the tools of the underwriter's trade today would have baffled his or her predecessors in 1960. They would, for example, have been totally unfamiliar with such terms as "percentage of completion," "overbillings," "compilation statements," and "deferred taxes." They would have been as wiling to rely upon a financial statement prepared by the contractor on the surety company's standard form as upon a certified statement from an independent certified public accountant. Sadly, while the underwriting profession has become much more sophisticated since then, the fundamental objective of an underwriting profit, seems, also, to have largely faded into obscurity through sizable contractor defaults.

Surety underwriters must be something of "jacks-of-all-trades." They must have extensive knowledge of construction accounting, finely honed analytical skills, the ability to understand and interpret a wide variety of contracts, comprehend many forms of loan agreements and lending practices, and acquire an intimate familiarity with construction case law and legal precedents. They must also possess a knowledge of construction itself and the ability to identify hazards peculiar to the bonding of each type of contractual commitment. They must then combine this potpourri of knowledge and skills with experience, sound judgment, and a healthy perspective to produce the desired results—of an underwriting profit (where net earned premiums exceed losses and expenses).

A BRIEF HISTORY

In response to a U.S. Congressional inquiry during the latter 1960s into the mandated and costly need for suretyship on federal contracts, the General Accounting Office (GAO) conducted an extensive audit of the surety's role, not only in qualifying a contractor, but in defraying losses to the taxpayers through completion of defaulted contracts as well. This inquiry was conducted over a period of several years, during which most of the underwriting and claim files of many sureties were examined, and the companies' top officials interrogated at length. Upon completion, the GAO submitted its highly favorable report on the surety industry to Congress. Their conclusions were entirely supportive of the need for continued bonding of all federal projects.

While the subject of surety bonding, or suretyship, is little known to the general public, it is an ancient profession dating back to biblical times and has always been fraught with perils. From wise King Solomon's warning that, "He that is surety for a stranger shall smart for it" to the forfeiture of Antonio's bond to Shylock in Shakespeare's *Merchant of Venice,* history abounds with evidence of the pitfalls awaiting the unwary surety provider. The surety industry has borne witness to these grim portents so many years ago, as it has "smarted for it" on frequent occasions since the advent of corporate suretyship around the turn of the twentieth century.

Some of the early pioneering companies of the surety business have abandoned the field altogether, and many of the remaining players have had to implement many significant organizational and underwriting reforms in an effort to return their operations to the profit column. This is not a business for the untried novice who seeks opportunity with a new set of liberalized underwriting standards. Failure upon failure by some standard, substandard, or secondary markets attests to the fact that no provider of surety credit can make radical departures from well established, tried, and proven underwriting principles and survive for very long.

One of the two earliest statutory requirements for the bonding of public construction contracts date back to 1894 when the U.S. Government passed the Heard Act, requiring bonding of all federal work. This was later superseded in 1935 by passage of the Miller Act, which, in part, established the rights of recovery under labor and material (payment) bonds by subcontractors and material suppliers. Subsequently many states have incorporated these provisions for eligible claimants under their statutory payment bonds; these state regulations are frequently referred to as "Little Miller Acts."

SURETY VERSUS INSURANCE

Though most surety bonding operations are an extension of the product lines offered by many of our nation's major multiple line insurance companies, there is little congruity in the types of indemnity provided in each case. Further distinctions will be found in the number of parties to the contracts, the recourse, or subrogation rights available following a loss, and means by which each may, and may not, defray its losses through third party indemnity.

Exhibit I-1 provides a cursory contrast between property and casualty insurance and surety bonding, as well as a brief glossary of surety terms.

THE COMMERCIAL BANKING CONNECTION

As will be discussed further along, in principle and practice surety bonding is much more akin to commercial banking loan practices. In Maryland, however, a precedent was established in the early part of the twentieth century when the state ruled that a bank could not endanger its customers' deposits by assuming the combined risks of extending both monetary and surety credit.

At the conclusion of each chapter there are questions and answers, which some readers may find helpful in testing themselves in reviewing the subject matter. In other cases, for purposes of "in house" group study sessions, the answers may be withheld and the participants tested. In particular, for the novice in accounting, the case study exercises on financial reporting and analytical procedures should prove very helpful, and serve as a valuable reference source.

EXHIBIT I-1   Surety versus Insurance—Summary of Contract Surety Distinctions Compared with Their Insurance Counterparts

Contract Surety	Insurance
Principal/obligor	Insurer
Surety	No counterpart
Penalty/penal sum	Limits of liability
Tripartite (three party) agreement	Two party agreement
Term of Obligation	
Indefinite until contractual obligation is fully performed	Specified policy period
What's Covered?	
Performance	Named perils
Right of Subrogation	
Surety against principal for loss paid to obligee, subcontractors, and suppliers	Against third parties only and their insurers in tort cases
Indemnity Available to Surety/Insurer for Loss Caused by Principal/Insured	
Officers, stockholders, affiliated partnerships, and corporations on behalf of principal	None

1

The Contract Bonds and the Parties Thereto

Following is the first case study involving the fictional general construction firm of XYZ Construction Co., Inc., domiciled in Philadelphia. Throughout this book the reader will come to know this firm, follow its fortunes, and examine its financial condition at critical junctures in its operations.

THE BID SECURITY

The City of Philadelphia has decided to let a contract for a new administration building and has advertised for bids, specifying in the invitation that bid security in the form of a 5% bid bond, cashier's or certified check will be required. XYZ decides to take out plans, and the estimating department proceeds to "take the job off" by preparing an estimate of all direct job-related costs, plus percentage factors for overhead and profit. XYZ finally arrives at a bid price and is optimistic that its price will be very competitive as there are only two other construction firms with plans—both of whom are known to have very full work loads already.

At this point XYZ would call its agent, advise him or her that the bid estimate will be $6,000,000, and request that a bid bond be procured. Once the bonding company's authorization was given, the agent would prepare the bond to be provided, which would describe XYZ as Principal and the bonding company as Surety, who are jointly and severally, "held and firmly bound unto the City of Philadelphia as Owner/Obligee in the full and just sum of either 5% of the amount bid, or expressed in dollar terms as $300,000." The condition of this obligation is that the contractor will enter into a signed contract for the amount bid, if low bidder, and furnish the required 100% Performance and Labor and Material, or Payment Bonds (the final

1

bonds) at that time (Note: Percentage bid bonds are commonplace today, and it is usually only in cases where the surety wants to hold the contractor's bid price strictly to the estimate, and provide no leeway for escalating last minute sub and material prices that would increase the estimate, that the surety will insist on a fixed dollar bid bond amount. This could perhaps connote some reservations on the part of the surety over its contractor client's marginal financial condition for carrying the aggregate work program that would result should the client be low bidder.)

By the bid stage the surety's underwriting should be fully completed and it would normally stand ready to furnish the final bonds when the contracts are signed. There could be complications after the bidding, however, that could cause concern to the surety, and these developments are covered in the following section on "The Performance Bond."

Should XYZ for any reason decide it does not want to enter into the contract after finding itself low bidder (maybe it found an error in its bid) it would—or should—promptly notify the City by telegram and request that it be excused from entering into the contract well before an award. If the City denies the request, some part, if not all, of the bid bond penalty may be demanded. In most bid bond default situations the contractor's and surety's liability would be the difference between the low bidder's price and that of the next responsible bidder. If this difference exceeds the bid bond penalty, there would be a "full penalty" loss. If XYZ fails to respond to its liability, the City would call upon the surety for payment. Upon making payment, the rights of the City would be subrogated to the surety, and the surety could then proceed to recover its loss from XYZ.

There is a rather uniformly observed practice followed by most surety companies regarding bid amounts. If the bid price exceeds the estimate given to the company by 10% it is expected that the agent or contractor will clear the higher amount with the surety before submitting the bid and bid bond. Unfortunately, with last minute prices furnished by some subs and suppliers (in some cases to avoid having their prices "shopped"), the contractor sometimes does not have time to notify the surety of the higher estimate. This is a recurring and common problem through the entire surety industry and does not normally result from having a deceptive or uncooperative contractor client. Infractions of this type are almost always treated on a case-by-case basis by the surety, with reactions ranging from a mild rebuke—or none at all in cases where the amount or percentage of the actual bid is only marginally in excess of the 10% guideline—to refusal to authorize the final bonds in much more severe cases where the excess may be substantially higher.

Sizable bid spreads are always a concern to the surety underwriter—and most likely to the low bidder as well. The customary measurement of a contractor's bid adequacy is another 10% industry guideline. If the second bidder's price is over 10% higher than that of the low bidder, or, in some cases where there is a 15 to 20% spread, for example, between the first and third bidders, there might be the suggestion that the low bidder had made an estimating error and would lose money if he or she proceeded with the contract. At very best these assumptions are only highly theoretical with little scientific merit. They are, however, the first yardstick by which an underwriter can evaluate the adequacy of the client's bid. In most of these cases, by the time the underwriter has had an opportunity to review the bids and question the contractor about the price, there has already been a thorough review of the bid conducted in the contractor's office immediately following their return from the bid letting. The contractor, therefore, should normally anticipate the underwriter's concern and be prepared to confirm that either they could find no error, or that there was one and they were taking immediate steps to be excused from accepting an award of the contract.

If the contractor confirmed that there was no mistake and the spread was very sizable, the surety may still want some corroborating documentation to support the contractor's contention. As examples, this support may take the form of a comparison of the client's unit prices with those of the other bidders, if they were available. Alternatively, it may involve having the subcontractors reconfirm their prices and provide evidence of bondability, if they had originally been required to do so before the bidding.

There are countless scenarios that can produce abnormal bid spreads, and very seldom are two alike. It is the underwriter's responsibility to probe into all possibilities, however, and form the best judgment possible based on the data available. Suffice it to say that the larger the bid and the greater the spread, the more intensive the inquiry should be at the earliest possible date.

Other forms of bid security may be required in the form of bid letters, consents of surety, or surety agreements. Under whatever styling, these undertakings commit the surety to provide the final bonds if its contractor is low bidder and awarded the contract. Unlike the bid bond, these agreements have no dollar limitation, and the contractor may or may not be joined as a party to this commitment. These "letters" may be issued as sole bid security or in tandem with a bid bond, and are considered by some surety underwriters as being tantamount to a 100% bid bond.

THE PERFORMANCE BOND

As it happened, XYZ was indeed the low bidder, and was moreover delighted with its price. It eagerly looked forward to an award by the City and the subsequent "Notice to Proceed." At the signing of the contract, XYZ must appear with the executed final bonds. The performance bond so tendered provides that the work will be fully and satisfactorily completed in accordance with the *contract documents,* which, by definition, include the Agreement Between the Contractor and Owner and all modifications issued subsequent thereto, the General Conditions, Specifications, and Drawings. The bond incorporates by reference all of these documents with which the contractor must conform. In the case of the Philadelphia contract 100% final bonds were required; however, in other cases something less than full penalties could be required. From a cost standpoint, nothing is gained by lower penalties, as the premium is based on the contract price—unless the penalties are so low that a specified maximum premium on the aggregate of both bonds is applied. In still rarer—if not now extinct—cases, there may be a combined performance and payment bond with a single penalty applying to both obligations.

As a practical matter, the underwriter cannot be expected to review all of the contract documents, or understand some of their technical content. It is principally the bid specifications he or she usually relies upon for contract details in cases of competitive bidding—particularly in connection with public work and private lettings, where, in the former case inalterable statutory forms are required, and with the latter, where standard A.I.A. documents are frequently used.

Most performance bonds include a provision requiring that the surety waive the right to be notified of any alteration or extension of time agreed upon by the owner and contractor. This would generally involve increases or decreases in the contract price and/or extensions of the completion date resulting from *change orders*. These changes usually develop when departures from the original plans and specifications occur as the work progresses. Substitution of materials, redesign of certain systems, additional work ordered by the owner, unavoidable delays not chargeable to the contractor, and so on, can be the basis for change orders. Regardless of the number of additive or deductive changes the performance bond penalty remains the same, unless, in rare cases, the surety signs a consent to the increase in its liability.

In other words, the surety has no control over any contract modifications the contractor and owner decide upon between themselves, unless the modifications are of such enormous magnitude that the "scope" of the contract is radically changed. This is largely a very gray and indefinable area, where ordinarily the surety's defense over what it considers totally unreasonable and unanticipated exposure is adjudicated in the hostile environment of a courtroom, or before an arbitration board.

On the other hand, in order to keep the surety fully committed to all change orders, some owners will require an executed consent of surety (not to be confused with same styling for bid letters) in each case, regardless of the size of the changes and notwithstanding the waiver provision in the bond form. Once the contract is fully completed the surety will adjust its original premium charge based on the final contract price. Should the final price be less than the original an "underrun" exists and a return premium is allowed. Should it be more, there is an "overrun" and an additional premium is charged.

LABOR AND MATERIAL OR PAYMENT BONDS

This is the companion to the performance bond. Unlike the performance bond, however, where right of recovery accrues only to the owner/obligee, and/or other dual obligees (the latter to be more fully described later in this chapter), the right of recovery under a payment bond usually extends to *all* unpaid claimants who are legally entitled to bring separate actions against the contractor and surety. A claimant under a general contract (as opposed to a subcontract) is usually defined as one having a direct contract with the general contractor (GC), or with a subcontractor of the GC for unpaid labor and/or materials. This would, therefore, provide recovery to a GC's subcontractors (first tier), their material suppliers and sub-subcontractors (second tier)—but *not* to suppliers to the sub-subs (third tier). State statutes usually govern in defining eligible claimants, except on federal work where the Miller Act controls.

Regardless of the number of claimants and amounts filed with the surety, payment bond liability cannot exceed the specified penalty. An exception would be where a claimant filed a mechanic's lien for nonpayment on a private job and the surety bonding the GC provided a *discharge of mechanic's lien bond,* effectively removing the lien from the building and transferring it to the bond (in Florida these are referred to as lien transfer bonds). The effect of the same surety furnishing both bonds is to increase its overall exposure to payment claims on a particular job (public work cannot be liened). In most states the lienor must "perfect" its lien (prove its grounds) in an equity court within one year for it to be considered valid and enforceable. While liens cannot be placed upon public works projects, they may be levied against contract balances.

Occasionally, only a payment bond will be required, for which there is a specific premium rating provision. Obviously, in these instances there are no performance guarantees, and it would seem that if the owner feels it necessary to have surety backing on payment obligations, one might well wonder if it is not being "penny wise and pound foolish" not to pay a little more for full protection.

MAINTENANCE BONDS AND EFFICIENCY GUARANTEES

Most contracts provide for a warranty or maintenance guarantee to cover defective or inferior materials and workmanship for a specific period from the date of acceptance by the owner—usually one year. Because this provision forms a part of the bonded contract and the contract, in its entirety, is "read into the bond," the surety and principal remain liable under the performance bond during whatever period is stipulated. Through the incorporation of this provision in the contract documents, therefore, separate maintenance bonds are rarely required where there is also a performance bond involved. Some owners, however, feel more comfortable in having a separate bond to specifically cover the warranty period and while there may be some redundancy in this requirement, the surety will generally oblige and furnish it for a one-year period without premium charge. After this period there is an annual charge based on the value of the guaranteed work. In unusual cases where a maintenance bond may be required in the absence of a performance and payment bond, a specific premium rating charge in the bond manual is applied.

Warranty provisions are not necessarily uniform as to all construction phases. Where, for example, a general one-year warranty might apply to all completed work there could be longer periods specified for particular phases (i.e., five years for roofing, three years for boilers, two years for landscaping). The underwriter must be satisfied that these longer term guarantees do not commit the principal and surety to inordinate periods of liability. A case in point would be in connection with roofing, where most sureties usually resist a five-year warranty, but may oblige their better roofing clients. Ten years, however, for anyone would most surely be considered prohibitive even though the roofer's commitment might be backed by a manufacturer's warranty as to defective materials (not labor) for an even longer period. Some of these manufacturers' warranties are "pass-throughs" directly to the owner.

It is important, also, for the underwriter to distinguish between warranties for defective materials and workmanship and *efficiency* guarantees. A guarantee that boilers would operate at designated pressure levels or that climate control systems would maintain room temperatures at specified degrees of Fahrenheit, for example, would be considered efficiency of operation guarantees. Where this exposure is contemplated, the underwriter must determine if the risk is an acceptable one (perhaps backed by a reputable manufacturer's pass-through warranty to the owner) and/or if this work has been subcontracted, that the subcontractor be required to furnish bond to the general contractor, so that the risk is transferred to the former's surety. Efficiency guarantees are specifically rated in the bond manual and, unlike maintenance guarantees, a premium charge does apply for the first twelve months.

SUBCONTRACT BONDS

Performance and payment bonds, required by general contractors from their subcontractors, guarantee that the subcontractor will faithfully perform the subcontract in accordance with its terms, and will pay bills for labor and material incurred in the prosecution of the subcontracted work.

The underwriting of subcontract bonds often presents more difficulties than writing bonds for prime contractors, for several reasons.

A subcontractor is more often than not awarded a contract without free and open competitive bidding. It therefore is difficult to know how his or her price compares with others. It depends largely on the job itself.

Also, the usual subcontractor—except those for excavation, foundations, and other preliminary services—is not called upon to perform his or her work until perhaps months after the start of construction; therefore he or she is buying for the future at the current cost levels.

It is frequently difficult for the subcontractor to mesh all his or her work to his or her own organization's benefit, and it may be that all or a large part of his or her subcontracts may come to a peak at one time, thereby presenting a financing problem. Labor and material costs can make or break any job, and delays and unfortunate staggering of subcontract work may speedily run up the overall costs, far beyond the earlier estimated point.

Certain types of subcontractors, on the other hand, present desirable underwriting qualifications, such as those of well capitalized and managed organizations. Prime contractors are usually willing and eager to seek them out because their work represents an important key in the entire job performance.

So far, in the case of XYZ, we have been thinking in terms of a general or other prime contractors having a *direct contractual relationship* with the owner. Most GC's will subcontract 50% or more of their work to subcontractors. Let's assume XYZ (GC) awarded the HVAC (heating, ventilating, and air conditioning) phase to

ABC Mechanical Contractors, Inc. (sub). ABC does not fabricate the sheet metal ducts needed for its subcontract and negotiates with RST Ducts, Inc. (sub-sub) to furnish these vital components. ABC's direct contractual relationship would be with XYZ and RST's with ABC. The owner would have no recourse against either ABC or RST, nor would XYZ have any against RST.

Some states, and other public and private owners, may bid and award the GC, electrical, HVAC, and so on, as *separate prime* contracts. While the GC may have the responsibility for coordinating the various specialty trades, it will have no direct contractual relationship with any of them nor any liability on their behalf. This arrangement can cost the GC the leverage it might otherwise enjoy against them as subcontractors—particularly in the area of directly controlling the "purse strings" and, where need be, of withholding payment because of poor performance and/or unpaid bills. In the case of separate contacts the owner has this prerogative and may or may not agree with the GC that payments to a particular sub should be withheld or worse, that it be declared in default.

The requirement by a GC that a subcontractor furnish bond is usually, but not always, left to his or her own discretion. There are cases where the GC's surety may require that the major subs be bonded, or in rarer cases, the owner may require it. Obviously, in the case of separate prime contracts, all bonds would run to the owner, as obligee, but as subcontracts the subcontractor performance and payment bonds follow the same format as those for the GC, except that the GC becomes the obligee.

Under the subcontract payment bond, claimant is defined as one having a direct contact with the principal—translated into its suppliers and sub-subs. Here again, payment bond liability stops at the second tier. In relatively few cases do we see a sub requiring a bond of a sub-sub.

DOUBLE BONDING

Sureties are usually very skeptical over furnishing a subcontract bond on behalf of one of their specialty subcontractor clients to one of their general contractors—even if the owner has not required bond of the GC. This could place the surety in the unenviable position of alienating and conceivably losing either or both of the accounts should a dispute arise between the GC and sub, and the former filed a claim under the subs bond.

In the event that the GC was required to furnish bond the surety's liability could be compounded if the owner subsequently made claim under the GC's bond, partially or entirely because of the bonded subs failure to perform, and the GC, in turn, made claim under the subs bond. Here, the surety could be liable under both bonds on the same job. As a general practice, most sureties will only consider double bonding situations if (1) both GC and sub are highly capable and valued clients, preferably with some experience in having worked together in the past and (2) the surety has the GC's concurrence in accepting its bond on behalf of the sub and the GC is aware of the implications in doing so. Should either the surety refuse to provide the sub bond, or the GC refuse to accept it, the sub would usually be forced to apply to another surety company, which could materialize into a permanent relationship because of most companies' aversion to becoming involved in "one shot" accommodations.

In order to assist its subcontractor client in such situations a surety may request another company to "front" for it and provide the sub bonds. This would generally be accomplished by the regular surety fully reinsuring the "fronting" surety without the latter having the benefit of full underwriting data on its temporary client. This is a highly questionable practice, where such subterfuge is often revealed if the GC should file a claim under the subs bond.

BONDING CONFIRMATION LETTERS

Occasions will arise when a surety or its agent (with surety's authorization) is requested to provide a letter confirming its intent to furnish performance and payment bonds for a specific contract on behalf of its contractor client at a future date, or one simply expressing general surety credit guidelines as to single job and aggregate program capacity, without reference to any particular contract. These letters are very similar to the bid letters or consents of surety discussed earlier under "The Bid Security," usually requested (1) by a private owner and/or his or her lender on behalf of a GC being considered for a project during preliminary negotiations or (2) by a general contractor whose policy may be to either selectively bond only certain subcontractors or all of them with prices in excess of a specified amount. The request for a letter from the GC develops when he or she is prequalifying a sub for bonding prior to submitting his or her own bid for the general contract, or subsequent to the bidding when, for a variety of very valid underwriting reasons, the sub is unable to furnish the bonds before execution of the subcontract agreement. For example, the surety may be requiring current financial statements, copies of the subcontract, work in progress schedules, and so on, before it is willing to accommodate its subcontractor client. In the meantime the GC seeks assurances from the surety that bonds will be forthcoming in a timely manner and may withhold payment of the sub's requisitions until they are received. Serious problems can develop for the GC when a subcontract is awarded based only on the sub's assurances that a bond will be furnished and such assurances bear little reality to an actual surety company commitment. Just as damaging to the GC would be where a surety company did provide an ambiguous letter with few, if any, conditions and later elected to "escape" from its misleading and suggestive language by refusing to provide the bonds.

In furnishing a letter regarding an established or prospective client to a third party, whether for a specific contract or a general statement of bonding eligibility, the most important criterion is *good faith* and to provide the recipient with a truly forthright and unambiguous expression of just precisely what the surety is prepared to provide and under what conditions. In cases where the contractor is well qualified to undertake the work and there are no underwriting contingencies or deficiencies, most letters will still be conditional, if only, for example, to require verification of a private owner's financing commitments and to review the final contract documents.

Timing between the date of the letter and the actual execution date of the bond may give rise to the need for conditioning the letter to provide for *contingencies,* particularly in cases where two or more general construction firms are being considered by a private owner for a contract that will not be started for many months from the date the letter is issued. The ultimate contract date may extend well beyond the period when, for example, new fiscal financial statements are due and the surety would have no way of determining what the contractor's uncompleted work program would be because of frequent bidding and negotiating activity for other new work in the interim. As a general rule the surety in these cases could not in good conscience provide an unconditional confirmation letter because of the uncertainty over future trends and developments. The surety's letter in this situation could very legitimately be conditioned upon their "satisfaction with underwriting considerations existing at the time of an award" or more positively, "barring any future developments of a detrimental nature." When such qualifications are preceded by the customary kudos regarding the surety's high esteem for the account, their enviable track record, the sterling character of the officers, and so on, there should be no chance of jeopardizing the contractor's negotiating position with the owner.

If there are underwriting *deficiencies* (i.e., excessive work programs, serious losses being sustained, unresolved claims by subs and suppliers, deteriorating pay record, default notices from owners, etc.), and the surety has not yet withdrawn from the account the wiser course for the surety would probably be to refrain from providing any type of letter that ethically should be conditioned upon resolution of matters better left undisclosed to a prospective owner or, in the case of a subcontractor, to the GC. To merely condition the letter as suggested for *contingencies* without further qualifying conditions would border on deception because of the negative implications in being able to provide a bond at all.

From moral, legal, and ethical standpoints, if a confirmation letter is to be issued by a surety it must not proffer unrealistic expectations and lull the recipient into the false sense that a bond will be routinely forthcoming—if, in fact, the surety has any reservations about that probability existing. To consciously or unwittingly convey unfounded "blue sky" inferences of bondability would be unconscionable and wholly devoid of professional ethics on the part of the surety and its agent participating in such duplicity. Moreover, legal action against the surety and perhaps the agent may well result through the owner or GC suffering monetary damages from his or her misplaced reliance on the letter's ambiguous language, which subsequently failed to materialize into a bond as inferred.

As a very recent case in point, the author participated as an expert witness on behalf of a large general contractor plaintiff in its suit against one of the nation's leading surety companies. This suit had as its foundation a letter the surety had authorized its agent to issue to a prospective subcontractor client, of whom the surety had little, if any, knowledge and no underwriting information. The agent and surety were aware that the letter was to be presented to the general contractor as evidence of bondability in connection with a substantial "fast track" dry wall subcontract for which the subcontractor was being considered. This was a case where the general contractor required that all subs provide bonds for the full amount of their subcontracts as a matter of customary practice.

In part, the letter simply stated that the agent "was in the process of establishing a bonding line with (company's name) in an amount sufficient to cover the entire subcontract"—with no conditions or qualifying comments. Because of subsequent underwriting developments of an unfavorable nature, the surety refused to authorize bonding of the full subcontract. The general contractor, in turn, withheld payment of several months' requisitions to the sub, causing him to abandon the job. To mitigate the time and expense of reletting the remaining portion of this defaulted subcontract, the general contractor undertook this completion with his own personnel, and brought suit against the surety for the additional cost incurred. After six or seven years of heavy legal expense, the surety finally settled out of court with the general contractor for a very considerable sum.

Where any question at all exists about the propriety in issuing a confirmation letter to begin with and the expression of actual intent the underwriter would be well advised to consult with his or her company's claims attorney or general counsel before proceeding.

SUBDIVISION BONDS

As a corollary to the bonds discussed in this chapter, the reader should be aware of the subdivision or off-site bond. This is an obligation, usually running in favor of a county or municipality as obligee on behalf of a developer/home builder principal, that indemnifies against the builder's failure to install whatever off-site improvements are required under local building codes—streets, curbs, gutters, sidewalks, and so on.

These improvements are financed and installed by the principal, and the bonds are normally required at the time building permits are issued. Sureties consider this a very hazardous obligation because of the highly speculative nature of real estate development and the burden of both the financing and performance responsibilities resting upon its principal. When authorized, collateral is generally required by the surety, as well as a bond from the contractor actually performing the work (if not the developer), which would run to the developer, as obligee. Where construction loans are involved, the surety may require that the cost of the improvements be escrowed or set aside for payment directly to the performing contractor in predetermined increments as various phases are completed.

"ARM'S LENGTH" DESIRABILITY

An additional risk is assumed by the surety when its bonded principal has a financial interest in the contract being performed. From an underwriting standpoint any involvement by a contractor with full or partial ownership interest in a project creates a duality of obligations not found in an "arm's length" contract, where the performance is fully paid for by an owner totally unrelated to the construction firm. In other words, the sole consideration for a contractor in an arm's length contract is to completely recover the direct job costs, plus a percentage of the general overhead burden and a net profit for his or her effort.

Where any element of an owner/builder exposure exists, the bonded principal must not only satisfactorily perform the construction phases, but be responsible for funding the entire project—the cost of which would well exceed the construction costs alone. The cost of the land, architect's and engineering fees, easements, and construction loan interest would be only some of the costs for which an owner/builder would be responsible in addition to the construction costs. The speculative element in most owner/builder projects is the real underwriter's concern.

Consider, for example, a real estate developer of residential property as a classic owner/builder case. The developer owns the land (usually along with the bank) and must pay for the off-site improvements and cost of the houses out of the construction loan. Unless there are "custom" homes, with a sales contract signed by a qualified buyer prior to construction, the houses are being built on a speculative basis. If these speculative houses don't sell after an extended period of time, the lender might well require that the construction loan balances be curtailed as a prelude to an eventual foreclosure. Assuming that the developer contracted with other firms to install the off-site improvements and couldn't pay them for their work as it progressed, they might very well walk off of the job site and file liens, which would have to be satisfied by the lender and/or the defaulted developer (if he or she was still solvent) in order to clear title for future sales of the unsold units.

The operative word in this illustration is *speculative*. Extending this hypothesis to a larger dimension, consider a case where the owner/builder has become the developer, constructing a 150,000 square foot auditorium. A "dummy" or "shell" construction firm has been formed, marginally capitalized at the surety's insistence and required to furnish a performance and payment bond running to the owner/builder and the lender as dual obligees (more on "dual obligees" ahead). The surety has taken the indemnity of the principals in the owner/building firm on behalf of its "shell" bond principal to provide underwriting substance. The surety felt comfortable with its position at first, but now it appears there will be a substantial cost overruns that will well exceed the original estimate and the lender's construction loan commitment. These overruns could result from a number of unforeseen developments chargeable to the owner/builder's contractual obligation—for example, correction of structural defects arising from faulty design, subsurface conditions not disclosed by the engineer's test borings, and so on—for which the owner/builder and its bonded "shell" corporation must bear the full burden of responsibility.

The lender in such a case might well suspend further loan draws for completion of the work, default its owner/builder borrower and the "shell" construction firm, and call upon the surety to take over the remainder of the work. Not only would the surety incur substantial completion costs, but it could also be besieged by a number of mechanic's liens filed by unpaid subs.

Had this been an arm's length contract, involving the same cost overruns, for which the contractor would not have been liable under the terms of its contract, the contractor could have defaulted the owner for nonpayment, pulled off the job, sued for balances still due, and been absolved from all further contractual responsibility,—along with relieving the surety's responsibility under its performance bond.

If the general contractor had entered into subcontracts containing "pay when paid" clauses, where permitted by state law, the subcontractors would have been unable to seek recourse for any unpaid balances due them, either against the general contractor or its surety—unless the GC had actually been paid for its work by the owner. (There is more on "pay when paid" provisions in the section on "Dual Obligees" below.)

These types of contracts involving commonality of financial interests between the contractor and owner are the type that must be underwritten with great care by the surety industry and considered above average risks. That is not to say they should not be written. Many are, and with highly profitable results when well structured by highly sophisticated investors. Conversely, they should only be favorably considered by underwriters with a great deal of experience and expertise in this field, and with reinsurers fully committed. For that reason most surety companies do not extend discretionary underwriting authority to their field offices for this class of business.

ADDITIONAL PARTIES TO THE BOND

So far we have been thinking of principal, surety, and obligee in the singular. As you will see below, there can be two or more in each instance.

Joint Ventures/Co-Principals

Two or more contractors may decide that because of job size, specialty trades, political reasons, and so on, it would be better to join together as one contracting entity and thus form a joint venture. They would enter into a joint venture agreement, which would name the parties thereto, describe the specific performance required of each partner, and set their limitations of liability, their respective percentage of participation in the profits or losses, their shares of any joint venture funding, the control of receipts and disbursements, the distributions of profits, and so on. With the joint venture formed and an awarded contract requiring bonding, each of the joint venture partners would be named as co-principals: "Jones Construction Co., Inc., Smith Electrical, Inc., and Johnson Mechanical, Ltd., a Joint Venture." In some cases the Surety Association of America's Form #1 (SAA Form #1) Application is executed by all of the partners, which specifies each one's limit of participation, the premium charge, the indemnification limit of each to the surety, and the amount of funding required by each in a special joint venture account. Usually, the surety will require unlimited indemnity from each partner for the full amount of the bond, notwithstanding their stated limitation of liability within the joint venture agreement.

The investment by a contractor in a joint venture should be set forth separately in his or her financial statement when it is considered material. When a contractor participates in a joint venture, further disclosure should normally be made in the financial statement, and this should include:

1. Share of joint venture earnings separately indicated in the income statement, and
2. A summary of the venture assets, liabilities, and operations set forth in a note to the financial statement, unless the venture is of such magnitude that separate financial statements should be included along with the contractor's statement.

The disclosure of this information is important, since a contractor participating in a joint venture will share in the loss as well as the profit up to the extent of his or her participation. Failure to account properly for joint venture operations could adversely affect a contractor's financial position without the immediate knowledge of the contract bond underwriter.

Silent Joint Venture

This entity is the same in principle as the "open" joint venture, except that one or more partners do not appear as parties to the bond or the contract. Their inclusion is accomplished by a separate silent joint venture agreement and execution of SAA Form #2 Application, which is essentially the same as Form #1. Most sureties question the joint and severability aspects of these agreements and treat them the same as indemnity agreements. With the wide use of general indemnity agreements today, which provide for joint venture participation, the use of Forms #1 and #2 is far less frequent than in the past.

Co-Sureties

In joint ventures each of the partners may enter into separate bonding arrangements covering their respective shares. For example, Jones, Smith, and Johnson each have different sureties, all of whom have agreed to participate in bonding their respective client's share. In joint venture nomenclature the partner having the largest share of the work will usually become the "sponsor" and his or her surety would become the "lead" surety. The sponsor would normally have overall control of the work, process all of the paperwork through its office, and have control of profit distributions to the others. The three sureties would sign the bond and limit their proportionate share of liability in the bond itself, or by a co-surety "side agreement."

There are rare cases, also, where two or more sureties would be serving one contractor, principally because of the very sizable contracts it undertakes, where no one surety's capacity would be adequate to cover the full liability, and/or to achieve a greater spread of the risk.

Dual Obligees

While there is usually only one owner, other interests can be named as dual obligees. The most common of these other interests would be a lender providing construction financing for a private contract. Other examples would include construction managers (CMs), developers, lessees, title companies, and so on. In each of these instances, the performance bond should contain a "savings" clause regarding payment to the contractor:

> There shall be no liability on the part of the principal or surety under this bond to obligees or either (any) of them, unless the obligees, or either (any) of them, *shall make payment to the principal, or to the surety* in case it arranges for completion of the contract upon default of the principal, strictly in accordance with the terms of said contract *as to payments,* and shall perform *all of the other obligations* required to be performed under said contract at the time and in the manner therein set forth.

Regardless of the number of dual obligees appearing in a bond, the limit of aggregate liability for the principal and surety is unchanged. However, the exposure to greater risk is assumed by virtue of the additional obligees, with dissimilar interest and commitments from that of the owner/obligee, afforded the right to seek recovery under the performance bond without being in privity of contract with the principal and surety—barring the latter's right of action against the dual obligees.

As the principal must perform—so must the obligee, and not only in terms of paying for the work as it progresses. Among other duties the owner is usually responsible for obtaining rights of way, zoning variances and, in some cases, certain licenses as well as the acts or omissions of other contractors in the case of separate prime contracts. Failure on the part of the obligee to perform could result in termination of the contract by the principal and/or denial of liability under the performance bond by both principal and surety. Failure to pay by the owner, however, could not be a valid defense against payment bond liability except in certain states that uphold the "pay when paid" clause (payment to subs conditioned upon owner's payment to the GC) incorporated into most subcontract agreements in use today.

QUESTIONS—CHAPTER 1

The Contract Bonds

1. What would normally be the surety's maximum liability on a percentage bid bond?

_____

_____

2. A surety provided separate 75% performance and payment bonds on a $1,000,000 contract that was declared in default by the owner, with a contract balance of $600,000 remaining to be paid. The price to complete the work from another

contractor was $750,000 and there were $300,000 in bills due subs and suppliers by the defaulting contractor. What were the surety's total costs to remedy the default?

_____

_____

3. Same question as #2 except that a 75% _combined_ performance and payment bond was furnished. What would then be the full cost to the surety?

_____

_____

4. Under a $3,000,000 bonded electrical subcontract, the firm furnishing and install-ing the switch gears for the subcontractor failed to pay $25,000 due its supplier for various components needed for the fully assembled units, and a claim was filed for the full amount with the surety. What was surety's liability?

_____

_____

5. What would your reaction be to bonding the following warranties under a general contract for construction of a fairground. If your reaction to the following state-ments is negative, place an "N" in the blank and give a short reason under each statement. If your reaction is positive, place a "P" in the blank and also give a short reason under each statement.

A. _____   The merry-go-round wouldn't break down for three months.

B. _____   There would be a 75% stand of grass within the first year.

C. _____   The paint would not peel off the auction barn for seven years.

D. _____   The drainage system would be adequate for maximum loads during the monsoon season.

E. _____   The roof on the cattle barn would not leak for one year.

6. Write a "T" if the statement is true and an "F" if the statement is false. Write the correct answer below each false statement.

A. _____   All underwriting should be fully completed before any perfor-mance or payment bonds are executed.

B. _____   The payment bond guarantees that there will be adequate funds available to pay the principal as work progresses.

C. _____   Payment and maintenance bonds can be furnished when no perfor-mance bond is required.

D. _____   100,000 square feet of four level underground parking decks topped by a 30,000 square foot health club can usually be added to a contract for a 50,000 square foot office building by change order without notice to, or consent from, surety.

    E. _____ In such a case the surety's liability would automatically increase by the additional cost of the work performed under the change order.

7. Indicate true or false to the following assertions, and provide the correct answers to the false statements.

    A. In an "up-front" joint venture between George's Plumbing Company with 10% participation, Harry's HVAC with 20% participation, and Pete's General Contractors, Inc., with 70% participation, Harry's liability would usually be limited to 20% of any loss suffered by the sureties by virtue of his obligation as a co-signatory to SAA Form #2.

    T _____ F _____

    _____

    _____

    B. In the foregoing question, if the same surety company executed the performance and payment bonds on behalf of George's Plumbing and Harry's HVAC, it would usually become the "lead" surety because of having bonded two participants in the three-way joint venture.

    T _____ F _____

    _____

    _____

    C. With the same three construction entities entering into a silent joint venture, the surety for the named party to the contract and bonded principal may have some reservations about the joint and several liability of the two silent partners executing SAA Form #1.

    T _____ F _____

    _____

    _____

    D. With a title company named as dual obligee on the performance bond for a general contractor, it would have recourse against the principal and surety through imperfections in the deed to the land upon which the building was constructed.

    T _____ F _____

    _____

    _____

    E. The construction loan lender and lessees of a manufacturing plant to be constructed would have identical interests in successful completion of the project with the owner and, therefore, be freely added as dual obligees to the bond by the surety.

    T _____ F _____

    _____

    _____

F. Failure on the part of the owner to provide payment to the contractor in accordance with the terms of the contract would enable the surety to terminate the contract.

T _____ F _____

_____

_____

ANSWERS—CHAPTER 1

1. Usually the difference between the low bid and that of the next responsible bidder.

2. $450,000. The difference between the $750,000 cost to complete and the contract balance of $600,000, or $150,000 under the performance bond and $300,000 under the payment bond.

3. The same. There was aggregate liability of $750,000.

4. None. This would have been a third tier claim. Surety was only liable to "one having a *direct* contract with the principal."

5. A. Negative. Efficiency guarantee.

B. Positive. A reasonable expectation for only one year.

C. Negative. Too long.

D. Negative. Efficiency guarantee.

E. Positive. A reasonable period for defective workmanship and materials.

6. A. False. Before the bid or private work negotiation.

B. False. Guarantees payment by principal to subs and suppliers.

C. True. Each is rated separately.

D. False. This would be considered a blatant scope change and the surety would have a strong defense in denying liability for the change order work.

E. False. The bond penalty remains the same regardless of the number of change orders, unless the surety signs a consent agreeing to an increase.

7. A. False. In most cases, the surety will want unlimited indemnity from all joint venture partners—notwithstanding their limit of liability in the joint venture agreement. With an "up-front" joint venture, SAA Form #1 would be signed.

B. False. Pete's General Contractors, Inc., had the largest share, and its surety would usually become the "lead."

C. False. The statement is true, except that SAA Form #2 would be signed.

D. False. It would be the owner who would have responsibility for clear title to the land.

E. False. Their interests would have dissimilarities, and the lender and lessee would not have privity of contract with the surety.

F. False. The contractor, as the party to the agreement with the owner, would have to serve termination notice.

2

Types of Construction Contracts

Contract underwriting requires a complete knowledge and understanding of—and satisfaction with—all of the contract documents that govern and control the contractor/surety liability for what can sometimes be extraordinary periods of time, particularly in the event of a default. For anyone who has had any exposure to construction claims, the importance of the contract language becomes abundantly clear, as it is argued by opposing counsels when defaults occur.

The forms of contractual agreements discussed below are the ones most commonly encountered.

STIPULATED SUM, LUMP SUM, OR FIXED PRICE CONTRACTS

(The terms are used interchangeably and are hereinafter referred to as lump sum.)

As XYZ Construction Company, Inc. prepared its bid to the City of Philadelphia, the estimator accumulated all of the anticipated direct construction costs and other job-related expenses, to which a gross profit and overhead factor was added. This comprised the total of the firm's lump sum bid, and the ultimate contract price after the award. As costs would vary during construction, XYZ's margin for overhead and profit would expand or constrict accordingly.

Following disclosure of its low bid, the "buyout" phase began, during which prices from subs and suppliers may have been negotiated downward until bare-bone costs were reached. By requiring fixed prices from its subs and suppliers, therefore, the element of uncertainty regarding future costs *seemingly* diminished, and the final estimate of total costs to be incurred became highly predictable. Should, however, XYZ elect to perform, for example, the masonry phase with its own forces instead of subbing or "brokering" the entire job, the element of uncertainty would theoretically increase, no longer insulated by a fixed price subcontract. XYZ would absorb the risks caused by weather, labor disputes, and so on. As just about every contractor has found, while uncertainty over cost variances can be reduced through transfer of risk to subcontractors (preferably giving bond to the GC), there can still be a multitude of complications that can seriously impact the predicted outcome of any job. Unbonded defaulting subs, strikes, disputes, weather, poor estimating, and so on, can all radically alter anticipated profit margins. Without regard to the circumstances, the contractor has no options for altering the lump sum price for his or her work or for having extensions of time approved, except through signed *change orders* from the owner and/or architect.

COST PLUS CONTRACTS

This type of contract is usually negotiated with an owner on either a *cost plus fixed fee* or *cost plus percentage basis*. In both cases, the contract delineates all reimbursable and nonreimbursable costs, as well as the fee or percentage of final costs to be paid incrementally as the work proceeds. There is less risk to the contractor in these cases, since regardless of cost escalations, he or she is still paid the same fee in the case of *fixed fee* contracts, and conceivably more than originally contemplated on *percentage* contracts, if the final costs exceed initial estimates. These are also commonly referred to as time and material (T&M) contracts.

COST PLUS FIXED FEE OR PERCENTAGE WITH A GUARANTEED MAXIMUM OR UPSET PRICE

When a "costs not to exceed" ceiling is incorporated, the *cost plus* contract bears a greater resemblance to a lump sum contract, and should be treated accordingly by the underwriter. With such an *upset* or *guaranteed maximum price* the contractor must use diligence in controlling costs, for obvious reasons.

In some *cost plus* contracts with a "ceiling," there can be a *savings clause* (not to be confused with the dual obligee savings clause), which provides for a split in the difference between final costs (plus the earned fee or percentage) and the *upset price*. Say, for example, this difference was $100,000; the contract could provide that the sharing between owner and contractor would be 75%–25%, 50%–50%, 65%–35%, and so on. Such savings could be generated by value engineering (cost saving recommendations advanced by the contractor), where these reductions might include alternate construction methods, design changes, or substitution of less expensive materials.

UNIT PRICE CONTRACTS

In most "heavy" construction involving excavation and grading (often as a part of general highway, street, or road projects), as well as oil, gas, sewer and water line contracts, prices are often bid or negotiated on a unit price basis (i.e., a price based on units of metric tons, cubic yards, linear feet, square feet, etc.). These unit prices are then multiplied by the number of estimated units required for a particular construction phase. The contractor's monthly requisitions would be based on the number of units in place during the most recent billing period, and payments to him or her would be made accordingly, once the architect or engineer certified the quantities claimed.

When "open-ended" unit price contracts without a guaranteed maximum price are entered into, they resemble a cost plus fixed fee contract, except payments would be based on unit prices instead of a breakdown of labor and material costs incurred, plus a pro rata portion of the fixed fee. There is, however, less margin for error in the unit price contracts, as labor and material costs plus profit are all compressed into the unit price—thus negating a recovery of unanticipated cost escalations.

As lump sum contracts, the maximum price would be established on the cumulative unit price totals for all phases of work to be performed. Change orders approved by the owner would be based on the unit price for any phase exceeding the specified quantity, including rock, if the contract contained a "rock clause." In the latter case, for example, it is very common to encounter substantial quantities of subsurface rock in highway and utility contracts, and, unless the specifications allow reimbursement of rock removal over and above the specified quantity, the contractor must bear the financial burden of such removal. In either case, "test borings" are usually made prior to the bid to determine what subsurface hazards do exist, including not only rock, but high water tables as well.

Some bids may include a combination of unit and fixed prices. This would be particularly true of highway contracts where clearing, grubbing, and mobilization costs would be priced on a lump sum basis, while most of the remaining would be unit price.

DESIGN/BUILD CONTRACTS

Design liability can be incorporated into any form of general contracts. The architect is usually the owner's agent with responsibility for monitoring construction progress, approving changes and monthly requisitions, earned estimates or progress billings. In the case of *design/build,* the *contractor* engages the architect and/or engineer as his or her agent—thus assuming full liability for adequacy of the drawings, compliance with building codes, cost of correcting design errors, structural deficiencies, and so on. In most cases, the architect will be independent of the construction firm, but under this agent arrangement, the construction firm can offer the owner something of a "package" deal. The architect does not generally appear as a party to the contract, resulting in the contractor assuming full contractual liability for both phases.

The *design/build* concept, therefore, with its additional hazards for the contractor, can present serious problems to the underwriter in deciding whether to get involved. One risk reduction solution would be to require that the architect, engineer, and/or contractor carry adequate limits of E&O (Errors and Omissions) insurance coverage and insert a disclaimer as to design liability in the performance bond, if possible. The latter would usually not be acceptable to the owner, and is, therefore, rarely accomplished because the owner's concern over design errors can be just as great as the surety's.

On the other hand, the design/build package concept can provide a smoother job for the contractor through diminishing some of the owner's checks and balances and ease in negotiating changes.

A higher premium rate contemplates the greater risk to the contractor and surety for this type of obligation.

TURNKEY CONTRACTS

While there are several variations of turnkey contracts, in its purest form, this can be considered the total packaging concept, where the contractor provides the land, drawings, and interim or construction financing. In these cases, title to building and land remain with the contractor until construction is fully completed—at which time the owner purchases the entire package through a buyout agreement.

The turnkey contract, obviously, encompasses many hazards. Untimely completion, funding complications, design errors, cost overruns, and so on, could negate the sale and leave the contractor with a fully mortgaged building he or she didn't need or want and/or conceivably put the surety in the real estate business if the contractor could not absorb the loss. Here, also, we could see something of a completion bond obligation to the construction lender, whose interests could be included as a dual obligee. The outright completion bond, usually running solely to a lender, is an obligation most sureties will not undertake regardless of the strength of their client, because the contract must be completed even though the lender may suspend funding of the construction loan. Because of the severity of the risk in these cases to the contractor and surety, specific premium rates also apply.

OWNER/BUILDER, LEASE BACK, AND PRIVATIZATION CONTRACTS

These contracts can take several forms with the ultimate objective in requiring a bond generally being to satisfy the interest of a lender and/or, in the case of subsidized low income housing projects, federal, state, county and municipal governments. One of the most common forms of this type of business enterprise is where a contractor/developer would form or syndicate a limited partnership, which would be the owner. The contractor/developer would usually appear as the general partner in the limited partnership, sell off limited partnership shares, and arrange for both construction and long-term or permanent financing. While the limited partnership would be the borrower and owner, the loans would be personally guaranteed by individuals of the construction firm whose interests in this conflicting role would be those of general partners. In other words, it would be nearly as though they were guaranteeing themselves (and the limited partners) that they would successfully complete the project, as well as pay themselves.

On bonds of this type, the contractor still appears as principal and the limited partnership appears as owner with the lender designated as dual obligee. The limited partnership/owner usually starts out as a "shell" with only minimal, if any, capital. Subscriptions to limited partnership shares are usually payable over extended periods of time and the limited partners themselves are only legally liable to the extent of their investment. The owner in these types of arrangements does not usually have sufficient resources initially to pay for any part of the construction costs and the contractor, therefore, relies solely upon construction loan draws by the limited partnership for payment.

Once completed, the limited partnership will lease its building and hope to achieve a level of occupancy that would provide a positive cash flow (income exceeding all expenses but depreciation), as well as the tax write-off benefits from depreciation. Under the publicly subsidized programs, interest rates are usually well below prime, with rent ceilings mandated by HUD or FHA at the federal level.

This form of contract had great appeal to limited partner investors prior to 1986 federal tax reform. The finished building was depreciated at an accelerated rate, and the individual write-offs could exceed investment by two, three, or five times. Because of the limitations on this form of deduction imposed by tax reform, the concept is far less popular today. Where the bountiful tax benefits were the principal attraction for investors before 1986, much more emphasis is now being placed on the economic viability of the project (i.e., positive cash flow returns and the potential for appreciation in value of the investment over the long term).

Lease-back contracts involve a similar form of owner/builder involvement. In these cases, the contractor will bid or negotiate the construction of a building with a prospective tenant who intends to occupy it on a long-term basis. This concept was at one time in wide use by the U.S. Post Office, which would take bids on their own plans and specifications for the construction of a postal facility. These bids would be based on five-, ten-, or twenty-year lease payments to the contractor (lessor), who would retain ownership and generally finance construction costs by means of construction and permanent loan arrangements. The lease payments would normally have been adequate to service the contractor's mortgage debt and leave an unencumbered building at the end of the lease term. Along the way, the lease income would be sheltered by depreciating the building, and if there were renewal options at the end of the initial lease term—and if they were exercised—the extended lease payments could all go into the contractor's pocket (less that due IRS if the building was fully written off by that time).

Lease-back contracts closely resemble "privatization" contracts, where the contractor not only finances the cost and retains ownership, but also *leases and operates the facility* as well. These contracts are transacted with public bodies, a case in point being where a municipality might require construction and operation of a wastewater treatment plant. Compensation to the contractor in the form of periodic lease payments should be adequate to cover debt service and operating costs, as well as providing positive cash flow.

Any form of contract involving the joining of interests of contractor and owner should be very carefully underwritten—or avoided in most cases. In general, it is desirable that there be an "arm's length" relationship between the two because of the additional risk inherent in speculative commercial building ownership.

CONSTRUCTION MANAGEMENT (CM) CONTRACTS

In these cases, an owner will usually engage a large general construction firm as his or her agent to supervise the entire project (similar to the architect's supervisory role, except far more extensive). There are many variations of the duties and responsibilities assumed by a CM, which can include (1) awarding subcontracts, (2) processing all job-related paperwork through its offices, (3) approving and paying monthly requisitions to the subcontractors, (4) negotiating and approving change orders, (5) settling disputes, and so on.

All of these functions are performed in concert with the owner and the latitude of direct control over a project by the CM is, of course, governed by the terms of the contract. While a guaranteed maximum price (GMP) by the CM may form a part of the contract with the owner, there are other cases where it does not, and the CM is merely paid a fee for its supervisory role.

THE SUBCONTRACT

The successful completion of many contracts has resulted from the inestimable contributions by the specialty construction trades. With the subcontracted phases of a general contract usually exceeding 50% of the contract price, the subcontractors can and often do have a vital impact on the final outcome. General contractors with a sound and mutually profitable relationship with their subs are often favored with better prices before the bidding than the GC who constantly shops the subcontractors' prices after a bidding, who fails to pay their monthly requisitions promptly, who holds up final payments and release of retention—or who fabricates bogus allegations for back charging the subs with delay and other spurious claims of little or no merit.

It is important to bear in mind that a subcontract bond does not mitigate deficiencies that may exist with a poorly performing subcontractor and that settlement of a GC's claim against a sub's surety may take months or years to settle if a dispute is the basis of the sub being defaulted.

In underwriting a subcontractor account for surety purposes the same investigatory procedures should be followed as with general or other prime contractors. In the case of subs, responses from GC's given as references, as well as those from architects, should provide excellent background intelligence. In analyzing financial statements particular attention should be paid as to how promptly the subs' accounts receivable are turning over, as well as to their trade relations with their suppliers.

Subcontract terms are another important underwriting consideration. Many GC's require that their own subcontract agreements be executed, which may be highly leveraged against the sub. The sub and his or her surety would be well advised to review these nonstandard subcontracts and bond forms carefully and to strike the objectionable terms altogether or negotiate less onerous terms with the GC. Actually, many subcontractors have wisely retained legal counsel for the purpose of reviewing subcontract agreements in every case.

THE GENERAL CONDITIONS

This is a document that delineates and defines the rights and responsibilities of the parties to the *Agreement,* as well as providing amplification of its various articles. For example, where the Agreement merely provides for the timing of payment to the contractor, the *General Conditions* specify the scheduling of values, the procedures covering application for payment, the architect's certification for payment, and so on.

In some contracts, hold harmless provisions relating to *Indemnification to the Owner* by the contractor and *Termination Rights by the Owner* should always be of interest to the underwriter. The two areas can impose onerous responsibilities upon the contractor through such *exculpatory clauses* as absolving the owner from responsibility for any of his or her own acts or omissions, as well as those of his or her agents and other prime contractors, transferring design liability to the contractor by holding him or her liable for not detecting the architect's mistakes, and imposing unreasonably short periods of time for the surety to respond to a default notice (to mention only a few).

While the Agreement between Owner and Contractor, therefore, relates specifically to a particular project, stating the contract sum, penalties, and/or liquidated damages, completion dates, and so on, the General Conditions are supplementary to the Agreement and can uniformly apply to many contracts with no changes.

In addition to the Agreement and General Conditions, the *Drawings, Specifications, Modifications,* and *Supplementary Conditions* also form a part of the construction contract.

In drafting general contracts, most owners have a strong propensity to "leverage" them against the contractor and thus his or her surety. It is, therefore, essential for both the contractor and surety to review the contract documents thoroughly before signing and, where possible, strike or negotiate any onerous terms. It is also incumbent upon the underwriter to thoroughly review nonstandard bond forms and, if they are unacceptable, to refrain from signing them until they are modified by changing or eliminating the objectionable sections or phrases.

All too frequently, bid and final bonds are executed without the underwriter having full knowledge of the contract terms or the content of nonstandard bond forms. Where such forms are required by an owner and there are questionable terms and conditions, the forms should be referred to the surety's claim department for interpretation and advice. There is usually very little, if anything, that can be done to alter statutory forms, but with private contracts and bond forms, the opportunity does exist—and it is in this area, as well as with subcontracts, that the underwriter and his or her client should be entirely satisfied that they are not being unfairly saddled with unacceptable risks and obligations.

At this point, it would also be well to emphasize the importance of confirming the private owner's ability to pay the full contract sum. Underfunded private owners have caused the construction and surety industries enormous losses through their inability to pay current billings and retainage to the contractor, thus affecting his or her ability to complete the work as well as pay subs and suppliers. Even in states where the "pay when paid" clause in a subcontract is upheld, the loss to the general contractor can be devastating.

It is sound business to require verification of the private owner's financing arrangements and to be totally satisfied that they are adequate. In most cases where there are questions about the owner diverting construction loan draws from one job to support another (which may, e.g., have incurred serious cost overruns), the contractor and surety should arrange for the lender to pay approved requisitions directly to the contractor, or to issue joint checks.

Extending the funding problem a step further, the contractor, surety, and owner should be satisfied with the solvency of the lender. His or her inability to fund the construction loan through insolvency can be just as damaging to all concerned as a financially distressed owner.

Where bid bonds are required for a private job and there is not absolute certainty about the owner's ability to fund full *development* costs (land, architect's fee, legal and engineering costs, and so on), additional wording should be typed on the bid bond that would condition execution of the contract and final bonds upon satisfactory evidence of full funding being furnished. Such evidence could be in the form of copies of the financing commitments and/or financial statements of the owner. Obviously, if an IBM, AT&T, or Chase Manhattan Bank is the private owner, this conditional language on the bid bond would seem superfluous.

Where no bid bond is required, the same precautions should be taken in the cases of bid or negotiated work well in advance of contract execution. Some contractors may be reluctant to question the owner's financial resources and, in these situations where bonding is required, the surety is usually perfectly willing to have the furnishing of financial evidence represented as its requirement (which it probably is anyway).

Of significant value to the contractor and surety alike are daily publications announcing new construction projects in the development stage, details of jobs to be bid and their bid dates, the prospective bidders who have taken out plans, and eventually bid results. One of the most prominent of such services is offered by the F. W. Dodge Division of McGraw-Hill Publishing Company—commonly referred to as Dodge Reports. These reports provide the names of owners, architects, and engineers, the bid security required, an estimated range of the bid amounts, the physical size of the project (square footage of buildings, linear feet of pipe, etc.) completion dates, and notice of awards.

They serve as a useful tool for the surety underwriter in providing advance notice of bid bonds that may be requested by his or her clients, as well as the opportunity to discuss specific job particulars with them well ahead of the bid date. If for some reason an underwriter has reservations about his or her client bidding a certain job, these can be discussed before the contractor incurs the expense of estimating the job. The reports further provide a good idea as to how intense the bidding may be by the number of other contractors with plans. The reports listing bid results can also alert the surety as to any abnormal bid spreads between the first three bidders and prompt an investigation of the spreads if their client is the low bidder. (There is more on bid spreads in Chapter 3.)

QUESTIONS—CHAPTER 2

Types of Construction Contracts

1. Under a cost plus percentage contract, where the contractor will receive a 6%fee on reimbursable costs incurred, subject to an upset price of $5,000,000 and a 50–50 split on savings with the owner, what would be the contractor's total gross profit with final costs of $4,300,000?

   _____

   _____

2. On a $15,000,000 private contract with a small owner/developer, the contractor was understandably concerned about the source of the owner's funding for the entire project. He or she was able to confirm that a reputable leader had committed on construction financing for the $15,000,000, as well as a permanent loan "take out" for the same amount. The contractor still wasn't satisfied on the funding issue. What other evidence did he or she require?

   _____

   _____

3. What is a completion bond and how are they generally regarded by the surety?

   _____

   _____

4. While E&O insurance could mitigate some of the risk in bonding a design/build contract, what very important feature would concern you most about this coverage, assuming the contractor's and architect's interests were named as coinsured?

   _____

   _____

5. In bonding construction of a speculative strip shopping center with a limited partnership "shell" as owner, and with the lender providing full financing of the total development costs to be named as dual obligee, you are also aware that the individuals in the construction firm to be bonded also appear as general

partners of the limited partnership. What reservations might you have?

_____

_____

6. What complications could arise from a turnkey contract if the architect's design error on a $3,000,000 project increased the construction cost estimate by 10%, and the lender refused to increase its loan commitment proportionately? Also, describe the dilemma for each party to the transaction.

_____

_____

7. If Phil's Construction Company entered into an "arm's length" stipulated sum contract with Able Developers, Ltd., and additional costs were incurred through the architect's design error, as agent for the owner, how would Phil's Construction Company mitigate its additional costs? How would the owner? How would the architect?

_____

_____

8. What would be the difference in the surety's risk in bonding a supply contract for the furnishing of cafeteria furniture to a school board versus one for supplying a data control system for a nuclear reactor plant?

_____

_____

ANSWERS—CHAPTER 2

1. $4,300,000 × 6% = $258,000 contractor's fee
   $4,300,000 + $258,000 = $4,558,000 total costs and fees
   $5,000,000 upset − $4,558,000 = $442,000 savings
   $442,000 divided by 2 = $221,000 contractor's 50% share
   $258,000 fee + $221,000 savings share = $479,000 gross profit

2. Source of the other nonconstruction development costs [i.e., land, architect, engineering and legal fees, right of ways and easements (if any), construction loan interest, etc.].

3. A bond running directly to a lender as obligee on behalf of a owner/developer. They are on the prohibited list of most sureties.

4. Limits, or amount of coverage. If, for example, there were structural deficiencies through design error and a collapse of an exterior wall, roof, floor, or parking deck occurred, there could be enormous personal injury and property damage

claims with the architect held ultimately liable. Compounding this with the time and expense of redesign and the construction costs of rebuilding, abnormally high limits of coverage might be insufficient in some of the more tragic cases. The size, location, and complexity of the structure would normally be the barometer for determining what might approach being adequate limits.

5. **A.** The individuals' dual roles in the construction firm and as general partners in a speculative venture.

   B. Having a limited partnership "shell" as an owner with its only capital being derived from payment of subscriptions by the limited partners over an extended period of time.

   C. The general partners' exposure as guarantors of the financing commitments, while also holding all of the stock in the bonded construction company. This could closely resemble a completion bond with the lender's interests named as dual obligee.

   In other words, the corporate and financial worth of the construction firm and its owners are being placed at grave risk through speculating on the economic viability of their venture—not just their success in profitability completing the construction phase for an adequately funded "arm's length" owner, the latter being considered risk aplenty by most sureties.

6. **A.** The contractor would have incurred additional costs, which must either be borne by him or her or recovered from the architect.

   B. The architect would be responsible to the contractor and bear the additional expense or make claim under his or her E&O policy—if he or she had one with adequate limits.

   C. If the additional costs could not be absorbed by the contractor and architect, the contractor might have to abandon the project—leaving the bank with a partially completed structure as its loan security.

   D. The surety may be liable to the lender for its loss if its interest were named as a dual obligee on the performance bond.

7. Phil's Construction Company submits a change order to the owner. The owner makes claim against the architect. The architect makes claim against his or her E&O policy.

8. There are numerous firms that manufacture a standardized line of cafeteria equipment, so that in the event of a default by the bonded principal, the surety could arrange for completion of the contract with relative ease. This would probably not be the case with a high-tech, state-of-the-art control data system—particularly if it was a patented process. The acquisition of patent rights and finding another firm capable of completing the work could be very costly to the surety.

3

Underwriting Fundamentals

There are relatively few contract surety underwriters who could build the most basic structure from a very elementary set of plans and specifications, much less a 50-story high-rise office building or a two-mile suspension bridge, or lay five miles of 60-inch sanitary sewer lines. From where then, with this lack of construction expertise, do underwriters summon the confidence to commit their companies to guaranteeing successful completion of literally billions of dollars in construction projects each year? What are the tools of their trade and how do they compensate for their lack of knowledge of the very product they are gambling so heavily on? This chapter will deal with those questions and attempt to describe the underwriting procedures that are designed to assure *no losses.* While this "no loss" hypothesis still remains an elusive—yet theoretically attainable—objective, most sureties faced with the harsh realities of highly unprofitable contract experience seek to achieve a loss ratio threshold that would still generate an underwriting profit after losses and expenses.

The foremost "broad brush" underwriting approach for both the surety underwriter and the commercial loan officer is satisfaction with their client's Character, Capacity, and Capital—better known as the Three Cs. When any one of these essential ingredients is in doubt, the underwriter should probably avoid the risk altogether. While there may be cases where, through "innovative" or "creative" underwriting, the lack of Capital or Capacity can be overcome by outside financial support, joint ventures, and so on, if Character is questionable, there is just no foundation at all to build upon.

How do you define "character" without being a psychoanalyst, clergyman, or Supreme Court Justice? In terms of evaluating the contractor risk, except through checking references thoroughly, there is just no easy or clear-cut way unless character flaws can be detected through such obvious misdeeds as a felony conviction for high treason or highway banditry. In the vast majority of cases, evaluation of character is a very subtle process which may or may not be accomplished by closely observing a client's work ethics, including, but not limited to, conduct during adversity and his or her history of honoring commitments. Many times the underwriter just has to rely on intuition and make a judgment call—which may, or may not, be entirely fair or accurate. The best course on this very nebulous point, barring the obvious, *is probably just to know with whom you are dealing and feel comfortable with them.*

Capacity would embody organizational depth, adequacy of plant and equipment, and experience. Would, for example, a general contractor specializing in fast food restaurants qualify to bid a 20-story curtain wall office tower? Has his or her estimating staff ever priced anything of this nature and magnitude before? Does he or she have an adequate Project Management team qualified to contend with the complexities of multistory construction? The same questions might be asked of a small water line contractor, experienced in three-foot trenching with a backhoe, who is interested in bidding a five-mile sanitary sewer line with 96-inch pipe. Does he or she have adequate expertise, equipment, and supervisory talent for such a project?

While the answers in these cases are very probably in the negative, they are extreme examples. Past experience in the line of work the contractor is proposing to undertake is a critical underwriting consideration. The substance, depth, and experience of key field personnel and adequacy of plant and equipment, as noted earlier, are the principal considerations in assessing the capacity of a contractor.

Job size alone is not the most significant consideration in evaluating the capability of a contractor to successfully perform a contract. It is important for the underwriter to look beyond the mere dollar size of a particular contract, compared to the largest one completed by the client in the past, and examine the construction characteristics and complexities of the new work to be performed.

The first underwriting standard would be the consistency of the type of construction involved in the proposed work with the contractor's past experience and areas of expertise. Consider, as an example, a contractor whose background includes having completed a $20,000,000 pre-engineered, three-story modular housing project; this contractor probably does not have the background to build a $15,000,000 three-story, site-built office complex—if his or her experience was principally in the modular construction field. Both contracts would be considered general contracts, but very different types.

Conversely, the modular contractor may well be qualified to undertake a $50,000,000 five-story housing project if it consists of preassembled modular units, if the contractor has adequate manufacturing facilities, financial resources, and field supervision, and if the work to be performed is within the geographical area of the contractor's normal operations.

If the same contractor was domiciled in Alabama and wanted to bid on a job of this size in Maine, the surety underwriter may have serious reservations about the logistical problem in transporting the assembled units over so great a distance, or the ability of the client to subcontract the manufacturing phase to a responsible local firm closer to the job site—not to mention the other impediments associated with taking work in unfamiliar areas of the country.

As another case study, consider the grading and paving contractor with a background of having completed a three-mile, two-lane section of county road as a prime contractor, at a price of $1,350,000. This firm now wants to submit a subcontract price to a large highway prime contractor for grading and paving a five-mile section of a six-lane interstate highway, estimated to run $5,000,000. While the first reaction of an underwriter might be to decline the request for bonding this subcontract phase as being too large in size and scope, compared with the largest single contract completed in the past, other considerations could alter this initial instinct. The contractor could well contend that he or she was entirely qualified, for the following reasons:

- The section of work would be within several miles of the contractor's office and yard.
- The terrain is flat and in open country, with minimal traffic control problems.
- The contractor's asphalt plant has the capacity to produce the quantity necessary.
- The contractor would only have to lease one additional grader and two dump trucks.
- There are plenty of qualified heavy equipment operators available.
- The completion time is more than adequate.
- The contractor's operations for the past ten years have been highly profitable.
- The contractor's working capital and net worth are sound enough to qualify for a larger aggregate work program than being considered in the present case.
- And the contractor has worked as a sub very successfully and harmoniously for the same prime contractor on the job to be bid over the past seven years on other smaller subcontracts, and knows the federal highway engineer who is to supervise this work very well.

These are all points that the underwriter should have investigated before making a decision based on the larger job size alone. Had the underwriter arbitrarily declined for only that reason, and disregarded the contractor's very valid reasoning, he or she may have missed an opportunity to generate substantial premium for the surety company through not having the subcontract awarded to their client. Moreover, the underwriter might also have given a more aggressive competitor an excellent opportunity to acquire the account.

This illustration should convey that, upon careful scrutiny of the salient job particulars, the prospects for successful completion of this subcontract were far more favorable than at first thought. Making informed decisions with all the facts before him or her is a paramount requisite for a seasoned and successful surety underwriter.

The capital factor is the one that most preoccupies the underwriter and consumes most of his or her time and mental dexterity. Here is where we determine the adequacy of a contractor's financial resources in terms of the job sizes and total work programs he or she wishes to undertake. (The terms "working capital" and "net worth" as used in this chapter should be interpreted by the reader as general overall financial strength in its broadest sense. Clarifications of each follow in Chapter 5.) To accomplish the financial evaluation, underwriters will analyze past and present financial statements of the business, work-in-progress schedules, personal statements, trade reports, long- and short-term credit availability, and so on.

Overextension of financial resources, coupled with shrinking or nonexistent profit margins, are the major causes of most contractor failures. Finances, however, in and of themselves, are far from being the only barometer by which a prospective account should be judged. It is only one side of the Three-C triangle, and without the other two sides firmly in place, a collapse might well follow. With this said, it is usually to the financial analysis that the underwriter gives first priority—and sometimes the last, when the statements don't measure up in either form or content.

Following is a hypothetical underwriting exercise, beginning with an application for bonding by XYZ Construction Co., Inc., after it became associated with EZ Bonding Company, and prior to the time the bidding of the Philadelphia Administration Building was authorized. The agent for EZ first called on XYZ at a time when XYZ's previous bonding company was barraging it with "too many questions," limiting its bonding capacity too severely, and charging excessively high premium rates. It was a very timely call, as XYZ had just that morning been declined on a bid bond needed for that afternoon. The EZ agent was given a warm reception and left XYZ's offices with (1) a completed questionnaire profiling the company's background, organization, references, continuity arrangements, financial reporting data, largest jobs completed, uncompleted work, and projections for future bonding needs, (2) last three corporate year-end statements, (3) personal statements from the officers, and (4) a letter from XYZ's bank describing available credit.

Happily, the agent was experienced and knew exactly what to ask for—thus making EZ Bonding's job easier and enabling it to expedite a decision.

While the EZ Bonding underwriter reviewed the accumulated data, a mercantile trade report was ordered, references were checked, and the statements analyzed. Upon completion of the underwriting procedure, everything looked very positive and EZ thought it had acquired a very strong and active bonding account—until it found that the next job to be bid was in Oregon for construction of a garbage recycling plant. As XYZ was domiciled in Philadelphia and had achieved its success as an office building/hotel general contractor, EZ Bonding balked at (1) XYZ venturing into unfamiliar territory 3,000 miles away and (2) undertaking a line of work which would have been a total departure from anything it had done in the past. Accordingly, EZ Bonding declined the Oregon job and while XYZ was disappointed, it nevertheless accepted the decision gracefully and returned the plans to the owner.

Shortly thereafter, the Philadelphia Administration Building job was put out for bidding and EZ Bonding authorized a $6,000,000 bid estimate, subject to XYZ requiring bonds from all of their subs with prices over $25,000. As XYZ was taking last minute sub and material prices, its management realized the bid was going to increase to $9 million. However, since they had a *percentage* bid bond (5% of the amount bid, instead of a fixed penalty of $300,000), they rationalized that, even though their bid price was well in excess of the estimate given EZ Bonding (i.e., 50% higher), they could justify the increase later and explain that there was just not enough time to have the higher prices authorized. In keeping with Murphy's Law, XYZ was the low bidder at $9 million, with the second bid at $10.5 million and the third at $10,650,000.

The next scene took place in the offices of EZ Bonding, where news of the bid results reaching the underwriter was received with grave concern. He or she asked for immediate explanations for both the higher bid price and the large spreads between XYZ's bid and those of the next two bidders. At this point, much thought was given to declining the final bonds; however, because of the $9 million bid, the percentage bid bond liability had increased to $450,000. Should this be forfeited, or should EZ take its chances on XYZ completing this contract within costs even though the next lowest bid was 16% higher? To further complicate matters, the City of Philadelphia decided to award the contract to XYZ the night of the bidding, and required that it be signed and the bonds furnished within three days.

EZ Bonding had to move quickly. It summoned the XYZ officers to EZ's head-quarters for an eleventh hour "what if" discussion. XYZ stoutly contended that it had reviewed its price several times and had a 6% gross profit margin "locked in." Its subs had reconfirmed their own prices and were anxious to furnish the subcontract bonds and proceed with the work. Furthermore, should EZ Bonding refuse to furnish the final bonds, XYZ would refuse to honor its obligation under the bid bond and bring suit against EZ for loss of anticipated profits and a host of other charges, including impugning its professional reputation, jeopardizing future bonding eligibility, and so on.

Clearly, a hostile relationship was developing between EZ and XYZ. Large sums of money were involved with any decision EZ made, and a meeting of EZ's entire Executive Staff was called into emergency session. A reevaluation of the entire account became the first order of business and the following facts were considered:

- XYZ had been in business for 15 years, and except for 1993, each year had been profitable.
- The largest previous contract previously completed by XYZ had been for $11.7 million, on which it had realized an 8% gross profit. This had been completed in 1996, and the project manager and field supervisor from that job would be assigned to the Philadelphia job—if XYZ could get the bond.
- EZ's antecedent investigation of XYZ had produced excellent references. Owners and architects on previously completed contracts had been highly complimentary and, in several cases, had provided very flattering letters of commendation. Without exception, every profitable contract since the company's inception had been completed in a timely and orderly manner. Dun & Bradstreet (D&B) and the National Association of Credit Management (NACM) reports reflected excellent trade payments to all vendors, and XYZ enjoyed one of their highest credit ratings. Each report also confirmed that there were no outstanding suits or judgments of record.

- XYZ's previous surety had confirmed a totally satisfactory relationship with it for many years. It had never received a claim from any of XYZ's subs, suppliers, or owners and, except for the declination of one bid bond request, it had followed XYZ on every other request during their long association.

  Note: For many years it had been the custom of EZ, when acquiring an account that had previously been bonded elsewhere, to routinely call the other surety company and inquire into its experience with the account, confirm the reasons given by the contractor for deciding to change (if he or she wasn't terminated by the former company), and confirm there were no claims outstanding.

  Because of numerous complaints from the subjects of these inquiries over alleged breach of confidentiality and damages sustained by derogatory, misleading, and inaccurate reporting of past experience, many surety companies have decided they will no longer participate in exchanging references on any of their accounts—past or present.

  Decisions to refrain from this practice also have some foundation in the U.S. Congressional efforts to modify or repeal the McCarron–Ferguson Act, which grants the insurance industry exemption from federal antitrust laws. The implication here could have been that two or more surety companies might have been acting in collusion to restrain the contractor from obtaining surety credit.

  Thorough investigation of a new account should negate the necessity for conducting prior surety checks altogether and avoid the legal implications that could result from a contractor's perception that it was maligned by an unfavorable reference furnished on its company.

- The major subs had worked very harmoniously with XYZ in the past and were all committed to going ahead on this project.

- XYZ's latest fiscal year-end audited statement of 12/31/98 reflected a sound, well-balanced financial condition with working capital and net worth amounting to 15% and 18%, respectively, of its total current work program of $21 million, including the Philadelphia job.

- EZ had the unlimited personal indemnity of XYZ's three officers and their wives. The officers' most recent personal statements, also, reflected sound financial condition, with good liquidity, and substantial real estate equity in all three cases.

- Continuity arrangements for the orderly transfer of each shareholder's stock in XYZ to the others in the event of their deaths, incapacity, or withdrawal from the corporation, had been provided for in a recently executed Buy/Sell or Shareholders Agreement. This agreement had been funded by life insurance policies with the limits of each adequate to purchase the full amount of a decedent's stock holdings from his or her estate.

- XYZ had established a $1 million unsecured credit line with one of the leading commercial banks, which had not been utilized in the past two years.

With all of these very favorable factors, EZ might have difficulty defending its decision to decline the final bonds, particularly with the already announced intent of XYZ to bring suit if it did. As concern mounted, the agent called with some very welcome news. He or she had just found, through an anonymous leak, that the second and third bidders had assumed, erroneously, that XYZ had decided not to bid, and as there were no other contractors with plans, they had collaborated to "sandbag" the City by loading their prices with profits well in excess of the norm. This gave immediate legitimacy to XYZ's price, and EZ Bonding authorized the final bonds with an admonishment to XYZ that, in the future, where a bid price exceeded 10% of the estimate given it initially, XYZ was to clear the higher amount with the EZ Bonding underwriter before submitting it. XYZ agreed, and the strained surety/contractor relationship became harmonious once again. As a fitting climax to this scenario, the conspiracy between the second and third bidder was leaked, also, to the City by a disgruntled estimator on the payroll of the third bidder and an indictment for bid collusion issued forth from the Commonwealth's Attorney's office.

While, of course, this is a fictional story, it has a great deal of basis in actual practice. Such things do happen occasionally, and when they do, they become very traumatizing for all the players. The events would have happened in the sequence described, and the review by the EZ Bonding executive staff did embody most of the underwriting considerations of sureties today. There are few, if any, construction firms that have been in business for several years or more that haven't had a bad job and suffered a loss of anticipated profit. It is the frequency, severity, and reasons for these losses that the underwriter must examine, ferret out their root causes, and be satisfied that the corrective measures implemented by the contractor will reasonably assure the avoidance of similar occurrences in the future.

The severity of one loss or frequency of several smaller ones has caused many outstanding construction firms to suffer an operating loss for the full fiscal year. Depending upon the circumstances of the losses, future prospects of a positive nature, and the general esteem in which the contractor is held by its surety company, there is generally no disruption in the surety/contractor relationship if the operating loss was an isolated incident, and the contractor had enjoyed many profitable years in the past.

It is when net operating losses occur back to back (two successive years), or when there are several losing operating cycles in the preceding five years, that the surety should seriously question its future with an account. Knowing when to get off of an account is every bit as important as knowing when to get on.

There is occasionally a paradox in deciding just how long to remain on an account when losses are being sustained and the bonding relationship would have been terminated except for a sizable bonded uncompleted work program outstanding. The concern in terminating the account might be severalfold, that is:

1. Denying the contractor the opportunity to acquire new and, optimistically, more profitable work could precipitate a collapse of the work underway, where anticipated profit margins were thin or nonexistent.
2. Once it became known in the "trade" that the account had been cut off, rumors could exacerbate the contractor's problems by, for example, having suppliers refuse further shipments, owners finding some pretext to withhold or delay

payment of current requisitions, and subcontractors walking off jobs if payments to them were not current.

3. The rumors could precipitate an avalanche of frivolous claims with the surety from worried subs and suppliers, as well as back charges from various owners who were anxious to serve adequate notice of anticipated problems and get on record in the event of the contractor's eventual collapse.

There are never two predicaments of this type exactly alike. The underwriter, with occasional input from his claims attorney, must make a critical judgment call in deciding the course to take, based on the circumstances peculiar to a specific case. While the old saws "The first loss is the best loss" or "Why throw good money after bad?" might well apply in most cases involving a troubled contractor, a little ingenuity in working with him or her on a tightly controlled basis could salvage a firm's capability to produce substantial premium income for the surety in its rehabilitated status for many years to come.

CREDIT FACILITIES

Having adequate credit facilities available to the contractor is important to the surety. It enables the contractor to meet payroll, pay subs, and retire trade obligations promptly during periods of cash flow interruption from "hung" receivables, sluggish inventory turnover, and so on. This becomes particularly important in cases where final payment (including retainage) are being unreasonably withheld, where there is a dispute between an owner and the contractor, and/or where there may be funding problems on a private job. For this and many other reasons, a short-term working capital line, *whether needed or not,* should be arranged.

Following is an examination of credit extensions in the forms most utilized by contractors:

Short Term

1. *Short-Term, Unsecured Credit Lines:* These loan arrangements enable a contractor to draw upon a line of credit *as needed,* up to a predetermined limit. Their purpose is to remedy temporary cash flow shortages, and repayment or curtailment of existing balances is usually expected by the lender on a short-term periodic basis as receivables are collected. While these credit lines carry no specific repayment or amortization schedule, many lenders require a 30- to 60-day "cleanup" period over a 12-month cycle—during which the contractor must repay the loan in full or be "off their books."

2. *Short-Term, Secured Credit Lines:* These are the same as the unsecured line except that the lender will require some measure of collateral to partially or fully secure the credit limit. This security interest can be taken on receivables, inventory, or fixed assets, and the security instruments (notes, real and chattel

mortgages, deeds of trust, etc.) are publicly filed in accordance with the Uniform Commercial Code (UCC). Most sureties object to the assignment of receivables and inventory because any impairment of these current assets could inhibit the contractor's progress in completing his or her bonded work.

Term Loans

These are loans with maturity dates in excess of 12 months and subject to repayment on a fixed monthly installment schedule. They are generally secured and used for the purpose of purchasing or borrowing against fixed assets (trucks, cranes, real estate, etc.). In these cases as well, UCC filings of the security instruments are made.

Revolving Loans

These are usually, but not always, secured loans that are long term in nature. As with short-term loans, there can be irregular borrowing and repayment activity, and often the pledge of receivables and inventory as collateral is involved. Some of the larger finance companies offer this form of credit for the purchase of heavy equipment, and the credit limit fluctuates periodically based on a specified percentage of the equipment's appraised value. Notwithstanding the collateral held to secure these and other credit lines, annual (or in some cases, more frequent) financial statements of the borrowers and guarantors are required.

Irrevocable Letters of Credit

As the term implies, these are irrevocable commitments by the issuing banks on behalf of their clients, providing sight draft authority to the holder upon the occurrence of specified events—generally failure to perform a specific agreement or legal obligation. While these credit instruments are usually taken by bonding companies as collateral on noncontract surety cases, they may occasionally surface on contract matters. Very rarely a general contractor will actually prefer these commitments from his or her subs in lieu of performance and payments bonds. In these cases the event that triggers the GC's right to draw down on the letters is failure to perform . and nonpayment to sub-subs and material suppliers. Irrevocable letters of credit are customarily used in lieu of surety bonds in other countries.

Loan agreements should be examined by the underwriter to determine if any of their affirmative or restrictive covenants impose unreasonable requirements upon the borrower and thus give the lender ground for accelerating the loan's maturity date and demanding payment in full. For example, certain covenants could require that unrealistic ratios and levels of working capital and net worth be maintained—the breach of which could enable the lender to "call the loan." This would obviously have dire consequences where such a loan default could result in not only loss of a credit facility, but seizure of the corporate assets securing the loan, as well.

On the positive side, there can be other covenants beneficial to the surety. The loan agreements can limit or restrict payment of dividends and purchase of fixed assets, or cap salaries and bonuses at a fixed amount—thus averting any undue dissipation of working capital. Lenders and surety companies have much commonality of interest in the success of their mutual client and should work in close harmony to foster that client's growth and prosperity.

SECURITY INSTRUMENTS

Indemnity Agreements

Except where applications are waived on some of the smaller, less hazardous noncontract surety cases, most sureties will require signed indemnification from the principal on their company forms. In contract surety, this is usually accomplished by the principal executing a general or blanket indemnity agreement—one that incorporates by reference any and all bonds issued on behalf of a contractor client without limitation of specific dollar liability.

If the principal is a corporation, the officer/stockholders and, in some cases, their spouses, may be required to sign as personal indemnitors on behalf of the corporation. The essence of their commitment is to make the surety whole on any losses it might sustain through providing bonds for the principal. In some cases, the surety may agree to place some dollar limitation on the amount of personal indemnification, or to the exemption of certain personal assets (i.e., homestead, etc.). Where there may be affiliated corporations, the corporate indemnity of one may be taken on behalf of the bonded corporate principal. In these situations, either an authorizing or ratifying resolution from the Board of Directors of the indemnifying corporation is required. Ratifying resolutions are given when the execution date on the indemnity agreement precedes the date of formal approval by the Board.

The supporting corporate, partnership, or personal indemnitor is not jointly and severally liable with the principal to the surety. If, however, these indemnitors appeared as co-principals and were named as parties to the contract and bond, they would be "up front" and considered jointly and severally liable.

Under the terms of some indemnity agreements demand may be made by the surety upon the principal and his or her indemnitor to post security when it is felt that the financial condition of the principal warrants additional strengthening. Also, under the terms of most agreements, the supporting indemnitors may notify the surety in writing of their intention to withdraw from future indemnification on behalf of the principal from and after a specified date. This notification is acknowledged by the surety in writing to the indemnitors, but their liability still remains for all uncompleted work prior to the date of release.

While most surety companies still occasionally use individual applications, with an indemnity agreement on each, taking signed applications on every bond is cumbersome and this practice has largely given way to the general indemnity form.

Collateral Agreements

When the surety feels its risk should be supported by collateral, it will require that assets, usually of a liquid nature, be pledged by the principal and that its own collateral agreement be executed. The assets most preferred would be certificates of deposit, savings account passbooks, marketable securities, and irrevocable letters of credit. In essence, the collateral agreement enables the surety to redeem the assets pledged for purposes of satisfying whatever loss it may have sustained.

Most *standard* surety companies will not entertain any prospective contract account whose acceptability is dependent upon the pledge of collateral. Some substandard or secondary surety markets will resort to collateral for this purpose and when they do, letters of credit are generally preferred.

Subordination Agreements

When the holder of a note from a contractor principal agrees to subordinate his or her interest to the surety, a subordination agreement is executed by the contractor (debtor), the note holder (creditor), and the surety. Essentially, it is agreed that no part of the indebtedness will be offered or accepted without the surety's written consent. This topic is discussed at greater length in Chapter 5.

BACKGROUND INVESTIGATIVE PROCEDURES (NEW ACCOUNTS)

Probably the most crucial caveat for any underwriter considering a new account is to *know with whom you are dealing!* Failure to fully investigate a contractor's background, as well as the officers personally, has been the cause of many contract losses, which may otherwise have been avoided. While the first impression of an account may be very favorable, there can be undisclosed "skeletons" in the background that, if known, could turn the surety's predilection for an account into an ultimate declination.

The more familiar an underwriter is with the contractors, architects, engineers, CPAs, banks, and public agencies in his or her region, the greater the probability that he or she has established "pipelines" for developing antecedent data that may not be a matter of written record or even circulated in the local "rumor mills." It has been the author's experience that phone calls to the reference sources provided by the contractor can often elicit more meaningful responses than the standard form letter inquiries used by most companies.

It is important to speak with someone in authority and with your assurances of confidentiality and anonymity make inquiries into specific facets of the responder's relationship with the subject of your investigation. For example:

For Subcontractors Contacted

- The number and amounts of subcontracts performed for the subject
- The subcontractor's opinion of the subject's scheduling of work and coordination of the specialty trades
- Manner of payment
- The subcontractor's opinion of the subject's job superintendents and project managers.

For Owners, Architects, and Engineers

- General opinion of job performance
- Opinion of subject's field organization—particularly as to outstanding individuals
- Disclosure of disputes and third-party claimants (subs, suppliers, etc.)
- Willingness to use again on other projects

For Material Suppliers

- High credit
- Selling terms
- Length of dealings
- Current balance
- Manner of payment

(In these cases, it is usually customary to speak to someone in the credit department.)

For Banks

- General regard for the account
- Length of dealings
- Range of current balances on deposit and past six months average
- Credit history
- Amounts and terms of existing credit lines
- Present loan balances outstanding
- Security required, if any (When accounts receivable and/or inventory are pledged, the terms relating to minimum working capital and net worth requirements, minimum working capital and maximum debt/equity ratios permitted and the negative covenants should be explored fully. In some cases, it could be important to have a copy of the loan agreement.)
- Third-party guarantees required
- "Cleanup" periods required, if any

A great deal of this information should be confirmed in writing by the bank.

When telephone inquiries are conducted, they should be done so in a congenial and informal manner. The more relaxed and casual the discussions, the greater the possibility of eliciting valuable information of a sensitive nature that might not otherwise be committed to writing. The inquiring underwriter should be particularly alert to the names of other potential sources for investigation during these discussions, names that were not provided as references by the subject. As these other sources are disclosed, they should also be contacted and queried.

Another undisclosed reference source may be the owners, architects, and subcontractors involved with a contractor's *work in progress at the most recent statement date.* The status of present work being performed is vital to the underwriter, since disguised or undivulged job problems may be the very reason the contractor is seeking another surety market.

As mentioned earlier, Dun & Bradstreet (D&B) and NACM reports are usually ordered by the surety when a new account is first submitted. D&B provides valuable antecedent, organizational and payment record data, together with notices of bankruptcies, judgments, and financing statements of record. Their graduated rating scale, similar to those used by A. M. Best Company and Standard & Poor's Corporation, is based principally on the subject's financial size and stability, and their ratings are widely held as reasonably reliable indicators of a firm's creditworthiness. NACM reports deal mainly with payment records and are far less comprehensive than D&B Reports. Despite the value of each of these reports to the surety, they should never stand alone as the sole source for background intelligence—rather, more as a supplement to the surety's own independent investigation, conducted along the lines described above.

THE CONTRACTOR'S MANAGEMENT TEAM

Full awareness and complete satisfaction with the personal background of the officers, directors, and stockholders of a corporate entity, as mentioned earlier, is the most critical element in contract underwriting. It is these individuals who will control and direct corporate destiny, and investigation of their ability, past affiliations, and reputation becomes one of the first orders in evaluating a new account.

Following the procedures described in the "Background Investigation" section, a thorough research of these individuals' contributions to the success or failure in their present position, as well as past performance with other companies for whom they previously worked, should also be conducted. Any personal investments of a speculative nature should be questioned in order to determine if an individual has a propensity for diverting corporate capital in the bonded entity to support such other ventures. Any suits, judgments, or liens should be fully explained to the complete satisfaction of the underwriter.

As previously covered, one of the most expedient means of forming a good personal profile of the individuals is to speak to owners, architects, material suppliers, and subcontractors who have been associated with the individuals in the past—being particularly alert to any suggestion of character flaws. Each individual's function within the corporation should also be examined. Whether they are principally occupied with internal management and financial affairs, or outside in project management, assessment of their proficiency and value to the organization within the sphere of their expertise should be evaluated in terms of (1) the significance of the vacuum that would result from their absence, and (2) who within the company could adequately fill this vacuum.

In terms of stock ownership, it would be essential to determine what disposition would be made of their stock holdings in the event of their resignation, termination, death, or incapacity. With two or more stockholders there should be a Shareholders, or Buy–Sell Agreement, under which the book value of the corporation is established and an option made available to each stockholder to purchase the shares of any other who, for any reason, had decided to divest his or her stock holdings and retire from the business. Ideally, these agreements should be funded by life insurance with the corporation as beneficiary in an amount adequate to cover the book value of the shares held by each.

In the absence of well-conceived continuity plans, a decedent's stock holdings would become an asset of his or her estate and purchasing the decedent's stock from the estate at book value, if his or her will so provided, might impose a severe burden on the corporation and/or the surviving stockholders. Alternatively, the surviving stockholders might be faced with holding a combined minority stock interest, and have control of the corporation rest with inexperienced and disinterested heirs, who may try to sell the decedent's stock back to the corporation for well in excess of the actual book value.

CORPORATE CONSIDERATIONS

A corporate profile should focus on the following points.

Length of Time in Business

While this aspect has been briefly discussed earlier, the greater the history of operations, the better the underwriter will be able to assess the trends developed under present management, that is, their track record in job performance and in meeting supplier and subcontractor obligations, how the officers and key personnel function as a team, how consistent they are in achieving their revenue and earnings objectives, and their commitment to corporate growth through the retention of net earnings.

Adverse characteristics also reveal themselves over extended periods of time. A spotty payment record, erratic gross profit margins, a poor record of completing jobs in a timely manner, diversion of working capital into nonoperational ventures, or an uncontrolled appetite for fixed asset expansion and heavy personal borrowings by the officers from the corporation are only a few examples of the other trends that become apparent in being able to review a contractor's operation over many years.

Divisions, Subsidiaries, and Affiliates

Where there are other business entities associated with the construction corporation (not necessarily corporations themselves), a full disclosure of their operations and financial condition should be required by the underwriter.

This is particularly important if they are divisions or subsidiaries of the parent corporation and their statements are not consolidated with its interim and fiscal statements. In the cases of affiliates, where there is full or partial ownership by the officers of the principal corporation, but no ownership by the corporation itself, full disclosure of the affiliates' financial condition nevertheless may also be required, particularly if the percentage of personal ownership by the corporate officers is a majority one, and if there are any business transactions, including loans or guarantees, past or present, existing between the affiliates and the principal corporation.

It is also important for the underwriter to determine the extent to which a subsidiary is dependent upon its parent for intercompany advances and loan guarantees (or vice versa), as well as any work performed by one for the other—resulting in what could be sizable intercompany accounts receivable. An audited consolidated and consolidating fiscal financial statement (more fully explained ahead) would divulge and eliminate all intercompany transactions. There are many variations of parent/subsidiary relationships, so that in each case the underwriter will be required to decide to what extent the subsidiary's operations impact the parent's financial condition.

In many cases, the corporate indemnity of a wholly owned subsidiary may be required on behalf of the parent corporation, supported by an authorizing resolution from the board of directors of the subsidiary. In cases where officers of the principal corporation own 100% of an incorporated affiliate, the underwriter may also require its indemnity, particularly where the officers do not personally indemnify the bonded principal. With partially owned affiliates, their indemnity may be a matter of negotiation, or may not be available at all because of the other unrelated interests involved.

NEWLY FORMED CORPORATIONS

Some surety companies will not entertain any construction firm unless they have been in business for at least three to five years. While others may be more flexible, they are still highly selective and apply abnormally stringent underwriting standards to those new organizations offering early potential of success in the near term of two to three years. Because of the initial organizational expense and the uncertainty in the procurement of new contracts, the underwriter can at best anticipate breakeven operations during the first year.

Under ideal conditions the incorporators of the new venture will have strong executive level backgrounds for many years with another well-established and reputable construction firm, with whom they enjoyed the favor of corporate management, and from which they resigned in good graces. In such cases it is not unusual for one or more supervisory personnel to also leave with the organizers of the new firm.

The new corporation's owners should be able to initially capitalize their corporation in an amount adequate to satisfy the minimum working capital and net worth requirements of the surety markets they intend to approach for a bonding relationship, and still retain a reasonably strong personal net worth. In some cases they may be able to attract additional capital from a third party investor, familiar with construction risks and rewards.

The professional venture capitalist, however, may not be the best source for raising additional capital, as such investors seldom have an appreciation of the finer nuances of construction challenges, would probably insist upon controlling stock interest, might expect a seat on the board of directors, and would resist providing personal indemnity to the surety. In this latter case, the requirement for personal indemnity from all officers should be considered a foregone conclusion—at least for the first several years, if not indefinitely.

There is no prototype of a new contracting venture that would appeal to most surety markets, where the background of the incorporators was questionable, and/or where there was a dearth of investment capital or a lack of strong construction experience. At best they may find a substandard, high risk market surety that would entertain bonding a few smaller contracts with some form of collateral, and that would charge abnormally high premium rates. The topic "SBA Surety Bond Guarantee Program" to follow will elaborate on this federally subsidized assistance for otherwise ineligible contractors to obtain surety credit.

OTHER UNDERWRITING CONSIDERATIONS

Why do contractors change surety companies? When presented with a potential new account the first question an underwriter will ask is: Why does this client want to change its present surety affiliation? This should not suggest an automatic negative response to an agent's submission on the part of the underwriter, but rather a more healthy reaction of being pleased that his or her company has been considered as a new market, and of wanting to know what attracted the contractor to this surety. In many cases it will be the underwriter who has seized the initiative and aggressively pursued a target account with an agent—rather than simply sitting in his or her office waiting for business to wander in.

As previously discussed, an agent must be constantly alert to competition for his or her accounts and be aware of the underwriting and marketing distinctions among the surety companies in the region. For the more common reasons listed below, an agent might counsel a contractor client on what other companies may be able to offer and recommend testing a new market if there is any dissatisfaction with the present surety, and if one or more of the following conditions exist:

- Inadequate bonding capacity with the backlog or work program limitations being imposed
- Refusal to consider bonding a particularly large or complex contract
- Premium rates too high
- Dislike of present company personnel (constant barrage of redundant questions and negative attitudes)
- Refusal to follow the contractor's operation into geographical areas unfamiliar to the contractor
- Refusal to release or modify requirements for third party indemnity

- Refusal to consider bonding the contractor on types of construction alien to the contractor's past experience

In the cases where a previous surety has terminated its relationship with a contractor, the following list of reasons might be found:

- Depletion of working capital and net worth through unprofitable operations and diversion of capital into unrelated ventures
- Numerous claims from subcontractors and suppliers and/or claims and default notices from owners
- Excessive fixed asset investment and corresponding debt increase
- Bad faith in withholding adverse developments and/or misrepresentation of financial data and operating trends
- Death or incapacity of a sole stockholder, with no continuity plans or key personnel to successfully conduct the business
- Inadequate accounting records, resulting in unreliable financial statements and work in progress schedules
- Arrogance in dealing with a surety underwriter and failure to cooperate by furnishing timely underwriting data

For these and other reasons a contractor may be attempting to arrange for a new surety market, and it is, therefore, of paramount importance for an underwriter to be completely satisfied with the contractor's reasons for wanting to make a change.

CLASS UNDERWRITING

While there may be certain classes of the construction trades that, either by their statistical failure rate as a group or by their general reputation, are among the less desirable surety risks, there are nonetheless many firms in these classes that are not only creditworthy in the standard markets, but that may become highly esteemed accounts. Such firms can produce substantial premium income and be considered target accounts by their surety's competition.

Unless a particular class is on a surety's prohibited list, it is important to look beyond the construction class distinctions and underwrite each account on its own merits. At the same time, in recognizing that some classes do constitute a greater risk, presenting more hazards, a surety needs to be extremely selective in choosing a client from such a class. Most surety companies strive to achieve a well balanced portfolio of contract business and welcome the opportunity to consider all accounts that have the potential of contributing to their profitable growth.

Of course, if a prospective account, regardless of class and financial substance, offers very infrequent potential for bonding activity, or appears to be a "one shot" deal, the underwriter must consider the issue of cost effectiveness; a great deal of time and expense is involved in the underwriting process, establishing a permanent file, and following for periodic updates on financial data.

THE HUMAN ELEMENT

One element that defies underwriting analysis is the impact that a personality conflict between a surety's bonding client and those parties with whom he or she contracts, or their representatives, can have on the final outcome of a contract. In some rare and unique cases a certain architect, owner, contracting officer or, in the cases of subcontractors, a general contractor may have—justly or unjustly—have earned the reputation for ruthlessness, unwarranted interference, or prejudicial treatment of a contractor under his or her direct supervision and control on the job site. In the vast majority of cases the underwriter can only rely on the contractor client's good judgment as to the type of treatment that client can expect when bidding or negotiating a contract. If it becomes evident that constant friction with the other contracting parties follows the surety's client from job to job, the strong assumption may be that the client is the one of a fractious nature, bringing about constant complaints to the surety's claim department.

Knowledge that a contractor has worked in harmony and profitably with an owner in the past, and has a well-established following of seasoned subcontractors, whom he or she has used frequently and with good result, is often an important underwriting consideration. The same reassurance would result from a subcontractor client having a history of profitable and harmonious dealings with various general contractors on previous jobs.

As a final adjunct to these underwriting observations, it would be hard to overemphasize the importance of an agent's acumen in presenting a new account to a surety, and of the company's branch office being thorough and professional in submitting the account information to its regional or home office. Many accounts have been declined because the local agent and company underwriter neglected to develop full information on the contractor, provided inadequate documentation or fragmented presentations, and failed to make well-founded recommendations as to the basis upon which an account should or should not be authorized.

HAZARDOUS CONTRACTS

Most sureties will generally avoid any type of contract involving removal of, or exposure to, toxic waste materials. With this exposure, the contractor's liability could be difficult to insure and very uncertain, both as to amount of potential damages and the length of time within which claims could be made. One of the most common types of this exposure would be asbestos abatement contracts, and many renovation jobs today do involve some degree of asbestos removal.

Contracts involving state-of-the-art technology and/or materials are another form of project about which sureties are often leery. The work is usually highly specialized in nature, often requiring the use of patented processes or systems. It is, also, not unusual to encounter long-term efficiency guarantees and very heavy research and development expenses. The risk of underwriting these types of projects is further compounded by the fact that finding a completing contractor in the event of a default could be difficult and very costly to the surety.

As mentioned in Chapter 2, owner/builder, turnkey, design/build and lease back contracts will also fall within the hazardous category.

SURETY MARKETS

The Standard Market

The larger "standard" surety markets are usually members or subscribers to the Surety Association of America, referred to in Chapter 1. This organization not only promulgates fidelity and surety rates, but is constantly involved in industry-wide monitoring of statutory bond forms, drafting of standard industry forms, lobbying for or against legislation at the federal and state levels, as well as serving on many construction and accounting industry advisory committees.

Deviations from the SAA, or "Conference" rates are more the rule then the exception today. Most sureties have filed their own deviated rates in the states where they are licensed, and many have structured these rates with two or more tiers—giving recognition to certain levels of net worth, audited statements, length of time in business, and so on. Where 15 to 20 years ago, such deviations would have been unthinkable by most stalwarts in the business, there are today only one or two holdouts still adhering strictly to SAA rates.

All of the standard—and most substandard—sureties are listed in the *Federal Register*—or what is more commonly referred to as the "Treasury" or "T List." Each company is listed with the limit it can commit itself to on any one *federal* job. This limit is based on 10% of the capital and surplus reflected on its most recent annual statement, with the changes made effective July 1 of each year.

Substandard or Secondary Markets

Substandard or secondary markets are usually more inclined to furnish bonds for contractors that cannot qualify in the standard market. These could include anything from the smaller emerging contractor whose financial worth does not meet the requirements of the standard markets to the large accounts with an unbalanced financial condition, losing operations, inadequately prepared financial statements, and so on. These smaller markets are usually inclined to require collateral to support their liability.

In many cases, their rates exceed those filed by the SAA, and their capacity to commit on any single bond is limited, compared to that otherwise available to a contractor qualifying in the standard market. While some of these sureties have prospered over the years, the failure rate among the group as a whole has been very high.

SBA SURETY BOND GUARANTEE PROGRAM

The ability of small contractors to compete with those of major stature was considerably reinforced by the Small Business Administration of the United States Government, when the SBA in 1971 set up a Surety Bond Guarantee program. This program guarantees bid, performance, and payment bonds required for a small business to bid on and carry out a contract, including but not limited to firms in construction, repair, maintenance, services, and supply.

This program provides a means, through an SBA guarantee, for small contractors to obtain the bonding required by obligees when these contractors are unable to obtain it through the normal bonding channels. These bonds protect the obligees from losses due to the contractor's failure to meet his or her obligations. The contractor remits a fee to both SBA and a surety company. The surety company issues the bond to the contractor and remits a portion of its premium to SBA. In the event the contractor fails to meet his or her obligations, the SBA will reimburse the surety company for a stipulated percentage of its financial losses. Nearly 80% of the firms receiving SBA assistance are in construction or related fields where bonding requirements are prevalent.

SBA's guarantee is at 90% for those losses incurred on contracts of less than $250,000 in face value, and is at 80% for losses incurred on contracts in the $250,000–$1,000,000 range.

SBA's guarantee of payment on losses can be extended to any bid, performance, or payment bond issued by a surety company on the U.S. Treasury Department's list of approved sureties. SBA does not issue bonds; it guarantees surety claim reimbursement of those issued by surety underwriters. Particulars of the Surety Bond Guarantee program are outlined below for the benefit of agents and brokers who, because of this facility, can secure bonds for small firms through regular underwriting channels.

Contracts of $1,250,000 or less are eligible for SBA's bond guarantee. If, however, a job has been put out to bid in components (e.g., site and foundation to one bid, building construction to another), and each does not exceed $1,250,000, each separate and identifiable contract can be guaranteed. There is no limit to the number of bonds that can be guaranteed to any one contractor.

SBA makes its own underwriting review and, if favorable, completes the guarantee agreement and returns it to the surety. As to the cost of the guarantee, each applicant pays a filing fee per bond. When the final bond is issued, the contractor pays SBA a percentage of the contract's face value (contract, not bond amount). When the bond is issued, the contractor will also pay the surety company's bonding fee or premium. The surety, in turn, will pay SBA a guarantee fee. SBA, in addition, will provide counseling and other supplementary assistance to the contractor in securing his or her bonds.

Participating sureties pay standard (as per individual agency agreement) rates of commission on these guaranteed bonds. There is a beckoning opportunity here for agents and brokers that should not be overlooked—that of assisting small contractors to get started and keep working, those who otherwise could not do so because their limited resources cannot meet normal surety bond requirements.

SBA also has a surety bonding line for more established and proven contractors. The bonding line cuts down on paperwork for all concerned, and should enable surety underwriters to make use of the SBA program more effectively than in the past.

A similar federal program, the 8A Federal Set-Aside, assists minority contractors by way of permitting contracts to be awarded to them on a nonbid "negotiated" basis. Once becoming "certified" under this program, the contractors generally enjoy greater gross profit margins than on competitively bid work, but only remain eligible for a period of seven years—after which it is assumed they would be able to effectively compete in the open market without assistance.

QUESTIONS—CHAPTER 3

Underwriting Fundamentals

1. If Jones Construction Company, Inc. was declined by its surety in requesting to bid a $10,000,000 suspension bridge over a river, on the grounds that the largest job completed by it in the past was a $1,500,000 supermarket, and the president offered to increase corporate capital by personally purchasing another $1,000,000 in capital stock to overcome the surety's concern, what do you suppose the reaction would be?

---

2. Same problem, except the president of Jones Construction Company, Inc. had persuaded Smith Mechanical Services, Inc. to participate with them as a joint venture partner. Smith was a very old, well-established, and strongly financed firm who had successfully completed a $28,000,000 HVAC contract on a 60-story office tower in the Midwest. What then might the surety's response be?

---

3. In considering an account for bonding, you, as the underwriter, find your applicant has established a $2,000,000 revolving line of credit with a large commercial bank which is secured by assignment of its accounts receivable. With $1,000,000 already drawn against the line, what would you suggest as a means of overcoming this objectionable feature, assuming all other aspects were favorable?

---

4. Charles Phillips, Dan Evans, and Chuck Davis each owned one-third of PED Construction Co., Inc. Phillips was 68 years old, Evans, 53, and Davis, 42. What would be your concerns, if any, if one of these individuals met an untimely death?

_____

_____

5. If you, as an underwriter for one of the major surety companies, were unable to approve an application for a small contractor because of its size and lack of track record, how might you be helpful in assisting this firm through either recommending or utilizing alternative solutions?

_____

_____

6. In approaching a prospective surety with a new account, what documentation should be furnished with the initial presentation?

_____

_____

7. Mark a "T" beside the following if the statement is true or an "F" if false:
   A. Co-sureties are jointly and severally liable for the full amount of a bond each executes.
   B. Each dual obligee is entitled to recover the full amount of the bond on which its interests are named.
   C. Failure of an owner to pay the contractor would provide a valid defense to contractor and surety for denial of liability under both the performance and payment bonds.
   D. Facultative reinsurance offers broader protection to the ceding surety than treaty reinsurance because separate acceptances are required in each case.
   E. Treaty reinsurers are historically tougher underwriters because of the periodic audits they perform on the companies' accounts.

ANSWERS—CHAPTER 3

1. Still decline. Money alone couldn't overcome the inexperience in completing a job so much larger, as well as compensating for undertaking construction of a type where they had no experience at all.

2. Decline again. How does completing a large mechanical contract qualify a contractor to build a bridge?

3. Ask the lender to furnish a written consent to the exemption of receivables from its loan agreement on bonded jobs.

4. Continuity (i.e., disposition of the decedent's stock holdings, which would revert to his or her estate). If there was no shareholders agreement or a "buy–sell" agreement funded by adequate limits of life insurance on each of the stockholders, the cost of acquiring his or her stock from the estate could create a financial hardship for the corporation if the estate, in fact, was willing to sell the stock back to either the corporation or the other two stockholders personally. You would also question what voids resulted in corporate management through loss of the services contributed by the decedent.

5. Either refer it to a company specializing in the placement of substandard accounts through the SBA Surety Bond Guarantee Program, or if your company was a participant in this program to accommodate the account subject to the 80% or 90% guarantee that would be provided—but this only where the prospects for successful completion of bonded work appeared reasonably certain.

6. Last three fiscal year-end statements, latest personal statements from the principals in the firm, full background through completion of a contractor's questionnaire form, letter from its bank confirming credit availability and the bonding needs.

7. **A.** True. Unless their limit of liability is specified either in the bond itself or by way of a co-surety side agreement among themselves.

   B. False. Only the bond penalty collectively.

   C. False. Payment bond liability to subs cannot be relieved except in those states upholding the pay when paid clause in a subcontract and then only where the language therein is very explicit on this point.

   D. False. The same.

   E. False. Treaty reinsurers are initially placed on risks by the surety without the benefit of underwriting information.

4

Introduction to and Sources of Financial Information

This chapter addresses the financial considerations most germane to the extension of surety credit, the formats used in financial presentations, the sources of those presentations, and examination of their depth and content. An overview of basic analytical disciplines is described, with a fuller explanation deferred to Chapter 8. These disciplines are briefly introduced in this chapter for the purpose of establishing a foundation for the more advanced treatment, and to set into context their relevance in the underwriting process.

The Introduction described how many of the principles of suretyship are analogous to the lending practices of a commercial bank. In both cases, evaluation of the applicants' creditworthiness is predicated upon very similar forms of financial information and analytical techniques. Requirements for surety undertakings arise from a multitude of very diverse statutory and nonstatutory (private) obligations, and often the ability to fulfill these obligations successfully hinges on the adequacy of the principal's capital resources. Supporting evidence of such adequacy, as required by the surety, is often found in the financial statements. Examination and analysis of the statements submitted, therefore, becomes one of the first orders of business in assessing the acceptability of a risk. Financial statements portray financial strengths and weaknesses as they are reflected in the value and composition of the assets owned, the amount, nature, and structure of corresponding liabilities, and the difference between the two. Upon evaluating these distinctions, the analyst should be able to make a well-informed judgment as to whether the capital required for successful performance of an obligation is available to the principal.

STATEMENT SOURCES

Implicit in the reliance upon a financial statement from any source is the fundamental premise that it has been prepared honestly and in good faith. Beyond that presumption, the user shoulder recognize that there are further distinctions to be drawn as to the qualitative reliability between one statement source and another.

Therefore, in relying upon a financial statement, it is important to know something about the individual or firm preparing it. This would embody knowledge of the accountant's academic and professional credentials, other major clients, reputation, the depth of the examination conducted, and the extent to which the accountant can reasonably attest to the material accuracy of the figures presented. Described below are various statement sources and how the surety might react to each one.

Internal or "In-House" Statements

These statements would be prepared from the principal's books and records by an officer or employee of the company, and range in reliability from the one prepared manually by a part-time bookkeeper of a smaller firm with few, if any, internal accounting controls, to the chief financial officer (CFO) of a much larger corporation, utilizing the most advanced state-of-the-art accounting and computer systems available. With certain exceptions discussed under noncontract surety, the underwriter in both of these cases would probably be unwilling to rely solely upon an internal statement for underwriting purposes, but would certainly give greater credence to the one more professionally and scientifically prepared, at least for purposes of interim financial reporting. The CFO for the large corporation may be a well-qualified certified public accountant (CPA), but one not independent of the company for which credit enhancement is one of his or her primary functions. Therefore, this could raise some question about possible conflicts of interest or objectivity in presenting his or her company's financial condition to the best advantage without implying impropriety or lack of professional integrity. Regardless of the confidence a surety may have in the quality and accuracy of internally generated figures, it will want corroboration from an independent accounting firm, whose job it is to challenge divergent viewpoints over treatment of certain statement classifications, the need for additional disclosures, and so on.

Registered Public Accountants

These accountants are licensed by the Department of Accountancy of each state. There have been no uniform educational standards established for eligibility, nor are the applicants usually required to pass a written examination. Because they are not held to the more exacting standards established for a CPA, they are not permitted to render any form of opinion or attestation on the financial statements they prepare. Accordingly, their accounting functions are usually limited to those of a bookkeeping service and, as a general rule, they are not acceptable as independent accountants for surety company clients.

The Certified Public Accountant

Most sureties will require that their clients engage not only a CPA, but often one who is highly qualified as a specialist in a particular accounting field. As no one doctor could specialize in the treatment of every malady and infirmity known to medical science, or any one attorney excel in all fields of jurisprudence, so could no one CPA be all things to all people in the many diverse fields of our business enterprise system. Depending on the accounting complexities of differing business organizations, there are varying degrees of specialized expertise required. As will be found in the treatment of contract surety later in this chapter, a high degree of specialization is required in the field of construction accounting.

To become certified, the accountant must not only pass the Uniform CPA Examination administered by the Board of Accountancy of each licensing state, but also have a minimum of several years accounting experience. The CPA is further held to a strict code of ethics and must adhere to highly principled standards of conduct, consistant with those established by the American Institute of Certified Public Accountants (AICPA), the dominant standard leader in the accounting profession. It is the AICPA, the Financial Accounting Standards Board (FASB), the U.S. Securities and Exchange Commission (SEC), and the American Accounting Association (AAA) that serve as the four independent authoritative bodies in establishing national accounting standards.*

It is, therefore, because of the magnitude of academic achievement the CPA must attain, the lofty standards established for his or her conduct, and the degree of accountability to which he or she is held, that the surety industry usually considers the services of a CPA essential in fulfilling their client's financial reporting requirements.

Many otherwise eligible contractors have failed to interest a prospective surety in their business by having inadequately prepared statements. After all, it is usually the statements that form the underwriter's initial impression of an account. Usually one can detect at a glance when a statement is professionally prepared or when it is confusing and incomplete as to accompanying exhibits, notes, and schedules.

The CPA's services, however, extend well beyond the mere preparation of a financial statement. The real professional will begin with recommending changes in internal controls and accounting methods, where needed, and counsel on the most advantageous financial and tax reporting systems. Recommendations may also be made for automating payroll and cost and revenue recording practices through use of the most suitable computer system, as well as regular visits by the CPA to the company's office to monitor implementation of any new procedures adopted. Tax planning is one of the CPA's most critical roles, together with his or her availability to intelligently discuss financial trends and developments with the users of the statements.

The CPA's Attestation   The extent of a CPA's examination of his or her clients' books and records is conducted in accordance with the generally accepted accounting principles (GAAP) established for the accounting profession. Three forms of attesta-

* See Bibliography.

tion may be expressed, each reflecting the scope and degree of thoroughness of the examination. In the ascending order of preference by the users, they are as follows.

Compilation   These statements are prepared from the contractor's books as kept by their bookkeeper. There are few, if any, adjustments and no accounting tests or verifications of assets and liabilities conducted. This is the least reliable of the three forms of attestation and rarely relied upon by grantors of bank or surety credit. Compilations do, however, organize and present a firm's capital resources and obligations in a professional accounting format and serve as useful management tools in tracking operating trends and current financial condition. They are also found revealing by credit grantors as mid-year or interim supplements to the more fully examined and reliable year-end fiscal statements upon which credit decisions are usually based. In some noncontract cases with minimal financial guarantee risk, they may be accepted as the sole underwriting determinant.

No opinion is expressed by the CPA as to the fairness and accuracy of the various statements. Because of the very limited scope of their examination, compilation statements are usually unacceptable for credit purposes.

Review   This form of examination does involve adjustments and accounting tests by the CPA. However, there is no direct confirmation of assets and liabilities with debtors or creditors, no physical inventory taken or, in the case of construction accounting, no visits to the job sites to verify costs incurred or cost to complete estimates. Sureties feel more comfortable with this degree of examination and usually, but not always, rely upon review statements for underwriting purposes.

While no opinion is expressed on review statements, the CPA may conclude the attestation by stating that "no discrepancies were discovered that would materially alter the representations of the accompanying financial statements."

Note: As there is no direct confirmation made of the accuracy of cash, cash equivalents, accounts receivable and payable, investment portfolios, and notes payable, the underwriter may require schedules of each and conduct his or her own independent verification.

Audit   This is a fully comprehensive examination using all of the generally accepted auditing procedures, including, but not limited to, observation of physical inventory, direct confirmation of bank balances, receivables, payables, and evaluation of cost and revenue estimates used in percentage of completion computations for construction firms.

The CPA will express an opinion on the fairness and material accuracy of the statements. If any facet of the statement does not conform with generally accepted accounting practices, a paragraph in the Report of Independent Certified Public Accountants will state specifically the nonconforming item(s) and the potential effect they may have on the financial statements.

In construction accounting, a highly accurate cost recording and job allocation system is essential in tracking line item costs versus the job budget on at least a weekly basis. As mentioned, one of the most important services the CPA can perform for a client is assisting in the selection and programming of a computer system most adaptable to the contractor's particular needs. This establishes an early warning system and enables the contractor to promptly identify and address unanticipated cost escalations, as well as producing reasonably reliable balance sheet and P&L statements periodically for internal use by the company. For greater comprehension of the accounting terminology to follow, reference is made to Exhibit 4-1, found at the conclusion of this chapter, and to Chapter 5, The Financial Statement.

Examples of CPA Reports

Audited Statement

Board of Directors
General Construction Company, Inc.
123 Main Street
Mid City, Arizona

Gentlemen,

We have examined the balance sheet of General Construction Company, Inc., as of June 30, 1991, and the related statement of income and retained earnings for the year then ended. Our examination was made in accordance with generally accepted auditing standards, and included such tests of the accounting records and such other auditing procedures as we considered necessary under the circumstances.

In our opinion, the accompanying balance sheet and statement of income and retained earnings present fairly the financial position of General Construction Company, Inc., at June 30, 1991, and the results of its operations for the year then ended, in conformity with generally accepted accounting principles applied on a basis consistent with that of the preceding year.

Very truly yours,

Review Statement

We have reviewed the accompanying Balance Sheet of the ABC Construction Company as of July 1, 1991, and related statements of income, retained earnings, and changes in financial position for the year then ended in accordance with the standards established by the American Institute of Certified Public Accountants. All information included in these financial statements is a representation of the management of ABC Construction Company.

A review consists principally of inquiries of company personnel and analytical procedures applied to financial data. It is substantially less in scope than an examination in accordance with generally accepted auditing standards, the objective of which is the expression of an opinion regarding the financial statements taken as a whole. Accordingly, we do not express such an opinion.

Based on our review, we are not aware of any material modifications that should be made to the accompanying financial statements in order for them to be in conformity with generally accepted accounting principles.

Compilation Statement

The accompanying Balance Sheet of the ABC Construction Company as of July 1, 1991, and the related statements of income, retained earnings, and changes in financial position for the year then ended have been compiled by us.

A Compilation is limited to the presentation in the form of financial statements information that is the representation of management. We have not audited or reviewed the accompanying financial statements and, accordingly, do not express an opinion or any other form of assurance on them.

CPA Preparation Methods   Examples of statements prepared with each of the following methods discussed are shown at the beginning of Chapter 8 as Exhibits 8-1 through 8-13.

All nonconstruction business organizations must choose between the cash and straight accrual methods of accounting for financial and tax reporting purposes. Paradoxically, in making this selection, there are cross purposes to be served. While income tax liability should be minimized by the method most conducive to the control of net income recognition, income should be maximized for financial reporting to a firm's credit grantors. As found from the following discussion, there are very distinct advantages and disadvantages in each case.

For the construction industry, there are two other accounting methods available, which are variations of the accrual method. These evolved from the elements of uncertainty over the ultimate cost of the contractor's finished product. To more equitably reflect the income actually earned on long-term contracts during each accounting period, either the percentage of completion or completed contract method addresses the erratic and unpredictable nature of cost variations throughout completion of a contract. With the completed contract method, there is no recognition of earnings generated from uncompleted contracts on its profit and loss statement, and the balance sheet provides specialized treatment of costs in excess of billings, and billings in excess of costs, on each uncompleted contract (per Exhibits 8-12 and 8-13). For construction firms, the cash and straight accrual methods are not acceptable under GAAP—for financial reporting purposes—but they remain acceptable for tax accounting.

Percentage of Completion Method   The percentage of completion method does give recognition to income generated from uncompleted contracts, but only as it is earned through application of a formula determining actual costs incurred to the total estimated costs upon completion. There is also balance sheet treatment given to the cost/billing relationship, similar to that used with the completed contract method, except earned income is added to the costs incurred and in arriving at earned revenues (per Exhibits 8-1 through 8-6).

Because of the complex and specialized nature of these two accounting methods, they are discussed in greater depth under the future topic, "Deferred Income Taxes" in Chapter 7. Limitations imposed on each of these methods by recent tax legislation are also covered.

Following are the major distinctions between the cash and straight accrual methods:

Cash Basis   (See Exhibits 8-7 and 8-8.) All accounts receivable, accounts payable, and accruals are eliminated from the balance sheet. The profit and loss statement (P&L) reflects only cash receipts as revenue, against which the cash paid for direct construction costs is deducted in arriving at a gross profit or loss. From this, the general and administrative (G&A) expenses incurred are subtracted, and other income received and other expenses paid are reflected. The remainder is net profit before taxes. The *modified* cash basis allows for depreciation and accrued taxes.

This method is usually unacceptable to any users of the statement for financial analysis purposes. It is impossible for the examiner of a cash basis statement to know the true financial condition of the company since accounts receivable and payable are not reflected. For tax reporting, however, this method is frequently used.

One of the reasons the cash basis is so popular for tax reporting is that receipts and disbursements of cash at the close of a fiscal year can be artfully controlled. Collection of receivables can be postponed to the next year, and liabilities can be prepaid as much as possible, thereby decreasing revenue and increasing expenses. These maneuvers can defer significant tax liability.

It should be noted, however, that the cash basis accounting is available for tax purposes only to companies with average annual receipts of $5,000,000 or less. Within this limitation, a firm may use the accrual method for financial reporting and the cash method for tax purposes.

Accrual Basis   (See Exhibits 8-9 and 8-11.) The accrual basis states receivables, payables, prepaid, and accrued expenses on the balance sheet, and revenues on the P&L consist of all amounts billed during the year, from which all direct job costs, including accounts payable and accrued expenses, are deducted to arrive at gross profit or loss.

Accrual is the basis most extensively used by the majority of business organizations today. As mentioned earlier, the completed contract and percentage of completion methods are themselves part of the accrual basis of accounting and are GAAP for the construction industry. As with the cash basis, taxable income under the accrual basis can be controlled by withholding billings at the end of a fiscal year.

As a general overview and summary of the cash and accrual methods, think first of all of the Federal tax returns most of us prepare each year. We show all of the income we have received from salaries, bonuses, commissions, dividends, interest, and so on, and reduce this by the itemized deductible expenses we have incurred (or the standard deduction). This is strictly a cash-in/cash-out declaration. If we are owed money by someone else, we don't have to declare this as a part of our adjusted gross income—nor can we apply the money we owe as a deductible expense. The same cash accounting principles apply to a business organization when reporting income on the cash basis. As amounts due and owing (accounts receivable and payable) are not factored into the recognition of income and expenses, they are similarly deleted from the company's balance sheet.

The obvious advantage and principal purpose, therefore, of utilizing the cash method is to mitigate or defer tax liability into future accounting periods. The disadvantage is that it serves little or no purpose in obtaining credit because of the elimination of vital financial disclosure on both the balance sheet and the profit and loss statement.

Conversely, the accrual method, as the term implies, requires that *accrued* accounts receivable and payable be factored into both the balance sheet and the P&L. Income recognition would include all amounts *billed* during the year as total revenues and direct costs or cost of revenues would include all cash paid plus *accrued* accounts payable (see the straight accrual statements of XYZ Construction Co., Inc., Exhibits 8-9 through 8-11, for further clarification). The same accrual principles would apply to other prepaid and accrued expenses reflected as assets and liabilities, respectively, on the balance sheet. The advantage of the accrual method is that it portrays a more accurate and usually stronger financial condition for credit enhancement purposes. On the downside, it is less conducive to control of taxable income.

EXHIBIT 4-1 Glossary of Accounting Terms

Accounting Period or Cycle   Any 12-month period selected by a company to record the results of its operations. This can be a calendar year or a 12-month period beginning with any other month of the year. Also known as a fiscal year.

Balance Sheet   See the section "Components of a Financial Statement" in Chapter 5.

Capitalization   All cash paid for the purchase of capital stock and any asset donated to a corporation, with the latter being classified as "paid-in capital—donation" and included as part of net worth. (See discussion of the corporation in Chapter 6.)

Cost of Revenues   See *Direct Costs* below.

Depreciation   Annual reductions to the cost of an asset to provide for wear, tear, age, and obsolescence. Reduces asset value on the balance sheet and earnings on the P&L.

Direct Costs   Cash expenditures and accrued liabilities for labor, material, subcontracts, and other costs directly associated with development of a product. Also referred to as "cost of revenues."

Earned Income or Profit   Earnings in excess of *Direct Costs*. Also known as *Gross Profit* and used principally in percentage of completion accounting.

Earned Revenues   See *Revenues* below.

Financial Reporting   Statements prepared for the benefit of financial markets, clients, stockholders, and regulatory bodies.

General and Administrative Expenses (G&A)   Unallocated indirect costs of operations (rent, telephone, etc.). Reduce gross profit on the P&L to net operating profit. Also referred to as "overhead" or "SG&A" (sales, general, and administrative) expenses.

EXHIBIT 4-1 *(Continued)*

Gross Profit   See *Earned Income or Profit* above. Total *Revenues* less *Direct Costs of Revenues*.

Income Recognition   Income or *Gross Profit* recorded during a fiscal year.

P&L (Profit and Loss Statement)   See the section "Components of a Financial Statement" in Chapter 5.

Revenues   Total gross receipts and/or accrued billings during a fixed year as described for the four accounting methods. Also referred to as *Earned Revenues* in percentage of completion accounting.

Treasury Stock   Outstanding capital stock repurchased by the corporation.

QUESTIONS—CHAPTER 4

Introduction to and Sources of Financial Information

1. Why is it important to know the sources of financial information? Explain how they would normally impact an underwriting decision.

2. What use might a surety make of an internal statement prepared by the CFO of a large construction firm? Of one prepared by an independent Registered Public Accountant?

3. What services are usually offered by a highly professional accounting firm, and why is each important to the surety?

4. When might a compilation statement be useful to the contractor? To the surety?

5. How might the surety develop greater confidence in the reliability of a review statement?

6. As a commercial loan officer for a bank, what supplemental information would you require in order to provide a more realistic financial picture when your applicant furnished you with a cash basis statement?

_____

_____

7. What single factor most distinguishes revenue recognition on an accrual statement from that prepared on a percentage of completion basis?

_____

_____

8. Explain how you would defer recognition of taxable earnings when using the cash and accrual accounting methods.

_____

_____

9. Why is it important for the accounting firm preparing a contractor's financial statement to be totally independent of its client's operation?

_____

_____

ANSWERS—CHAPTER 4

1. Satisfaction with the reliability of the information furnished—in terms of the preparer's knowledge, reputation, experience, and objectivity (the latter relating to the degree of independence from the firm being examined). Most sureties will, in part, base an underwriting decision on their comfort level with these four important characteristics.

2. For one prepared by the CFO, the surety would normally rely upon this for interim financial reporting purposes (mid-year statements), and as a measure of the client's accounting credibility when comparing its year-end statements with those prepared by its independent accounting firm. Little or no credence would be given to any statements prepared by a registered public accountant.

3. Counseling his or her client on internal accounting controls, automating payroll and cost recording practices, frequent monitoring of the firm's accounting department, and tax and financial planning.

4. The contractor would find them useful for internal financial management purposes, (i.e., current financial condition, trends, etc.). The surety may rely upon them for interim financial reporting.

5. Direct verification of bank balances, accounts receivable and payable, investment portfolios, and so on.

6. Schedules of accounts and notes receivable and payable, total billings and costs for the accounting period, and the applicant's most recent tax return.

7. Revenue for accrual would be total billings. For percentage of completion, only a percentage of the gross profit earned on uncompleted contracts plus corresponding costs would be recognized, in addition to gross profits on all completed contracts during the accounting period.

8. For cash, billings would be deferred at the close of their fiscal year, and accounts payable would be prepaid to the maximum extent. For accrual, billings for the last month or two prior to their fiscal year end would be withheld.

9. To avoid any appearance of conflicting interests through challenging divergent viewpoints over treatment of certain statement classifications, the need for additional disclosures, and so on.

5

The Financial Statement

COMPONENTS OF A FINANCIAL STATEMENT

A complete set of financial statements will include the following in the sequence in which they usually appear.

Report of Independent Certified Public Accountants

This will advise the form of attestation rendered by the CPA and either express an opinion, if an audit, or a disclaimer, if otherwise.

The Balance Sheet

Total assets, liabilities, and net worth, or equity, are stated. Illustrations of working capital and net worth computations appear on Exhibits 8-2 and 8-11. (This is discussed later in greater detail.)

The Profit and Loss, Income Statement, or Statement of Earnings

These terms are used interchangeably and hereafter referred to as P&L. In it total sales or revenues, expenses, and profit or loss are stated.

Reconciliation of Net Worth or Statement of Retained Earnings

This presents changes in capital stock and retained earnings not reflected on the P&L, for example, additional capitalization, payment of dividends, and so on.

Statement of Cash Flows

The basic purpose of a statement of cash flows is to provide information about the cash receipts and cash payments of a business during the accounting period. The statement is also intended to provide information about all the investing and financing activities of the company during this period. Thus a statement of cash flows should assist in assessing such factors as the company's ability to generate positive cash flows in future periods, to meet its obligations and the need for external financing, reasons for differences between the amount of net income and the related net cash flow from operations, the cash and noncash aspects of the company's investment and financing activities for the period and lastly, the causes of the change in the amount of cash shown in the balance sheet at the beginning and end of the period. The statement may be prepared using the direct or indirect methods.

Under the direct method, the income statement must be converted from the accrual basis to the cash basis. Changes in the balance sheet accounts that are related to items on the income statement must be considered. The accounts involved are all current assets or current liabilities.

Under the indirect method, certain adjustments are necessary to convert net income to cash flows from operating activities. First, changes that occurred in current accounts other than cash must be analyzed for their effects on cash. Then items, such as depreciation, that affected net income, but not cash, must be taken into account.

In either method, a reconciliation of net income and net cash flows from operating activities is provided.

Notes to the Financial Statement

These describe the accounting methods used by the company for financial and tax reporting, methods of depreciation of fixed assets, schedules of notes payable, details of over- and underbillings, contingent liabilities, subsequent events, pending litigation, loans to officers and affiliates, and any other disclosures of material consequence to the users.

SUPPLEMENTAL INFORMATION

There are also numerous other items of information that are not usually included in the body of the CPA prepared financial statement, but ones that the surety will find helpful. The well prepared client will have these schedules available on a regular basis. They are as follows.

Schedule of General and Administrative Expenses (G&A or Overhead)

If the P&L does not itemize G&A expenses, as on Exhibits 8-3 and 8-10, a separate schedule may be provided. This information enables the analyst to identify abnormal expenditures that may be inconsistent with prudent cost control and profit retention objectives of the company, for example, excessive officers' salaries and bonuses, travel and entertainment. Sizable bad debt expenses might suggest management deficiencies in establishing a sound credit policy and/or inadequacy of management's investigative procedures in screening the purchasers of their product on open account selling terms.

Excessive legal costs could indicate a firm with a propensity for litigiousness, or involvement in the costly defense of its own failure to adequately fulfill contractual agreements.

Schedules of Accounts Receivable and Payable

While these schedules do not ordinarily appear on an audited statement, they may occasionally appear on an unaudited or internally prepared statement. The surety may also want an "aging" of each account to determine (1) which receivables are not turning over, or not being collected in a timely manner and (2) which creditors are not being paid within their selling terms.

Schedules of Equipment

A breakdown of these assets may appear on the statement of a manufacturer or other business organizations requiring substantial capital investment in the equipment needed for development of its product. This information would be particularly appropriate for highway, bridge, water and sewer, grading, and so on, contractors, that is, those who require the use of large quantities of heavy equipment in the regular course of business. These schedules should describe each piece of equipment, its original costs, the depreciation charged against it, and the resultant net book value. (Additional detail concerning this aspect is included later.)

Schedules of Inventory

In the case of a manufacturer, this schedule would provide three components of his or her inventory: raw materials, goods in process, and finished goods. An imbalance of very large quantities of raw materials and finished goods to goods in process and/or a disproportionate ratio of inventory to other current assets could signal sluggish sales. In later treatment of inventory turnover, the significance of inventory to other current assets, and its relationship to costs of goods sold on the P&L, are discussed.

With a contractor, a schedule of inventory-work-in-progress may be prepared. This would allocate the cost of materials purchased for contracts under way, which may be stored on or off the job sites and not yet recognized as costs incurred. This subject is also elaborated upon as components of the balance sheet are discussed under the next topic.

Schedule of Unallocated Indirect Costs

This classification of expenses would include those not allocated as direct job costs to specific contracts. Examples would be shop expenses where various components are fabricated for units to be used on numerous jobs in progress, equipment repair charges, and so on.

To repeat, the foregoing schedules are not a part of the financial statement, but may be furnished as supplements. The schedules of completed and uncompleted jobs for construction firms are, however, usually included as a part of their financial statement. These list all of the contracts completed since the preceding fiscal year-end (hereafter referred to as FYE), giving final contract prices and gross profits realized (see Exhibit 8-4).

For uncompleted contracts, this schedule details adjusted contract prices, billings, costs, earned profits, estimated costs to complete, and over- or underbillings (see Exhibit 8-5). These are more fully discussed under construction accounting.

THE BALANCE SHEET

For both those who may possess a general knowledge of accounting and finance, and those who may not, the basic fundamentals of each, as treated herein, will serve the broader objective of forming distinctions between the more commonly applied cash and accrual methods (found principally in noncontract surety cases), and those peculiar to the specialized field of construction accounting. These distinctions will become more significant by comparing the straight accrual balance sheet, shown in Exhibit 8-9, with Exhibits 8-1 and 8-12, illustrating the completed contract and percentage of completion methods.

The stated net worth on the balance sheet should approximate the liquidated value of the company, if all of the assets were converted to cash as of the statement date and all liabilities were paid. The exception here would be intangible assets, which have no physical existence.

The financial condition represented on the balance sheet is not the same as it was the day before, nor will it be the same the day after. It is in a constant state of flux and can change significantly within short periods of time due to calculated, miscalculated, or totally unanticipated events. It is, therefore, essential for the analyst to assess the relative consistency reflected on a single balance sheet by comparing it with its antecedents prepared in the three to five prior years. It is only through establishing these comparative trends that greater confidence can be gained in the reliability of the most current statement.

By reference to the various balance sheet exhibits, it is obvious that the asset side on the left must always balance with the liabilities and equity on the right. More fundamentally, think of its composition in terms of a home you are purchasing. The value of a home (the asset) would appear on the left, with the value always being shown at cost. The mortgage (the liability) would be on the right, and the difference between the two, or equity (i.e., cash paid) appears under liabilities. The totals on the right-hand column must always equal or "balance" with the total asset values on the left. As the mortgage balance is reduced, the equity increases proportionately. To illustrate:

HOME—BALANCE SHEET
12/31/80

Assets		Liabilities	
Home	$100,000	Mortgage	$90,000
		Equity	10,000
Total assets	$100,000	Total liabilities/equity	$100,000

HOME—BALANCE SHEET
12/31/98

Assets		Liabilities	
Home	$100,000	Mortgage	$50,000
		Equity	50,000
Total assets	$100,000	Total liabilities/equity	$100,000

At this point, the home may have a market value of $300,000, but the asset still continues to be shown only at cost. The notes to the statement could disclose that the fair market value (FMV) established by a recent appraisal is much higher.

On the statement of a business entity, assets are segregated into those that are considered current and those of a fixed or long-term nature on the left side, while liabilities are grouped as current and long-term in the right-hand column. Current assets can be broadly defined as cash and other assets, that are capable of being converted into cash within a relatively short period without interfering with the normal operation of the business. This period is usually one year, but may be longer for businesses having an operating cycle in excess of one year. An example of this would be when inventory is sold and converted into accounts receivable. As these receivables are collected, the conversion to cash is made. Also included as a current or liquid asset are cash equivalents. These are short-term, highly liquid investments both readily convertible to cash and so near maturity that there is insignificant risk of change in value because of changes in interest rates. Equity securities (stocks) do not meet this definition. Investments acquired with remaining maturities of three months or less qualify for possible inclusion as cash equivalents. Any investments maturing after one year would be reflected as other or long-term assets and not considered current or liquid. On the liability side, all obligations due and payable within 12 months would be considered current.

Current Assets

A closer examination will now be made of each asset considered current.

Cash   There is no question about this being current as long as it is in U.S. currency and other accounts that have characteristics of demand deposits. It can be described as cash in bank or on hand.

Cash Equivalents Certificates of deposit, time deposits, money market certificates, and commercial paper with maturity dates of 90 days or less, all held for redemption within one year or as temporary investment of working capital.

Marketable Securities Values of marketable securities can usually be quickly determined by checking the most recent bid prices on the stock or bond exchanges where they were traded. The less susceptible they are to wide fluctuations in price, the stronger the indication as to their stability and soundness. The CPA will generally declare asset values at cost unless there is a permanent decrease in value. The value would be written down to market value and the decrease reflected in the current year's P&L. However, if given only on the basis of cost, there may be a note to the statement giving the latest market values if the decrease in value is considered temporary and is material.

Accounts and Notes Receivable Since accounts receivable are usually the largest current asset, close attention should be given to how promptly the receivables are turning over, or being collected. Many companies prepare a monthly aging showing those accounts that are current—usually less than 30 days old—and those in the 30–60, 60–90, 90–120 days and older categories. Anything of significance in the 90–120 days and older column should be questioned and, perhaps, disallowed as a current asset, particularly if a dispute exists with the debtor. A contractor's retainage would not be included in this aging.

The CPA will examine the collectibility of receivables and establish an allowance or reserve for those the company determines questionable. Wording on the balance sheet would usually describe this asset as an "account receivable, less allowance for doubtful accounts of X dollars." Once it has been decided that an account cannot be collected, either in part or in total, it will be written off the books as a bad debt. This charge-off would be an expense on the P&L, having the effect of reducing both net income and the tax liability during the year of removal.

By virtue of the many historic legal precedents establishing the surety's right to accounts receivable on a bonded contract due by an owner to a contractor declared in default, they should inure to the benefit of the surety as a matter of due course as it arranges for completion of the remaining work with another construction firm. This is often referred to as the doctrine of "equitable subrogation."

Occasionally, conflicts have arisen with a lender who has taken assignment of receivables as security for a loan and duly recorded this security interest in accordance with the Uniform Commercial Code. Where this duality of interests exists, and the surety is called upon to complete a defaulted contract, costly litigation may result as each party attempts to perfect its right to these assets in the courts. Despite well-established case law on this subject, the trial and appellate courts in Florida recently upheld the lender's right to assigned receivables—only to be overturned by the State Supreme Court on March 16, 1989 (*Barnett Bank of Marion County, NA v. Transamerica Insurance Company,* Case #72,531).

The appeal of the trial court's decision was closely followed by the surety industry. Had it not been reversed by the Supreme Court, the historical legal precedents, holding inviolate the surety's right to accrued receivables, would have been nullified and the cost of completing a defaulted contract could have risen significantly—particularly with a large accumulation of retained percentages. It would have established the lender's priority to these assets and denied the surety the remedy of defraying its cost to complete by not having them available as a part of the remaining unpaid contract balance due by the project owner (the loss to the surety being the difference between the contract balance and the cost to complete—plus unpaid labor and material claimants).

On the contractor's statement you may also find, in addition to accounts receivable, the corresponding classification of earned estimates and retainage on uncompleted contracts. Where this category is used, the amounts due on completed contracts are usually carried under the accounts receivable classification.

Earned estimate is a term that has always brought confusion to the minds of many agents, contractors, surety underwriters, and even accountants and bookkeepers not familiar with the construction industry. The thinking in some circles is that this is money that has been earned in the same sense as profit.

The current earned estimate is merely a figure based on the estimated amount of work completed within a certain time. Not all of the earned estimate may be recorded as income since the recording of income depends on the method of accounting used, which is discussed in Chapter 7.

Before we can understand what the earned estimate item represents, we must first know what is meant by retainage. Retainage or retention, in the construction industry, means holding back a part of the consideration (contract price) for performance of a contract, should the owner (or bonding company) need it to provide completion of the project. The amount of the retention varies but in the majority of cases it is 10%. This means that as the contractor submits his or her estimate each month the owner pays only 90% of the estimate submitted and withholds the other 10% until final completion of the contract. This payment method, known as progress payments, goes back many years and the original thought was that the 10% was the contractor's profit and he or she was not entitled to it until he or she completed the job. The days of the 10% profit largely are gone but the retention theory is still used.

As described above, it should be noted that the "conversion to cash within one year rule" does not necessarily apply to retainages. Retainages could be well over 120 days old and will be collected when the jobs they relate to are completed. On larger jobs with over one year's duration, retainages can accumulate for much longer periods and yet still be treated as current assets. Conversely, where a general contractor or other prime contractor has withheld retainage from their subcontractors, and these extend beyond one year, they are still treated as current liabilities.

Other receivables, as opposed to those arising from normal operations, usually represent such transactions as advances or loans to officers, employees, or affiliates, amounts due from the sale of fixed assets, and so on. While these may be classified as current assets, there is a general propensity in the underwriting community to reclassify most of the other receivables as noncurrent in the analysis of working capital. When, however, it can be established that some part of these have been collected subsequent to the statement date and the conversion to cash has materialized, they will be treated as current. Sureties must bear in mind that a CPA will not generally complete and release the financial statement for perhaps 60–90 days after the fiscal year end, during which period many changes and new developments may occur. While these changes may not necessarily be described as "subsequent events" in the notes to the statement, the facts can be known and related to the surety.

Notes receivables can arise from the same sources as accounts or other receivables, but instead of being carried as open accounts on the company's books, a promissory note has been required as evidence of the debt. These are usually interest-bearing instruments and can be either payable on demand or due at a fixed future date. If the term of the note is over a period exceeding one year, the CPA will classify whatever is due within the first 12 months as a current asset and the remainder as notes receivables, due after one year. This portion will be shown in other or noncurrent assets.

From an underwriting standpoint, notes receivables are usually classified by the underwriter as noncurrent, except those that have actually been paid since the statement date. Thus we have seen that while a CPA may quite properly classify certain assets as current on the balance sheet in accordance with generally accepted accounting principles, the underwriter may choose to treat them differently in arriving at what is considered to be those assets that will follow the 12-month evolution into cash with reasonable certainty.

Costs in Excess of Billings or Costs and Estimated Earnings in Excess of Billings

These are generally referred to as underbillings. This is an asset category peculiar to construction accounting and is treated separately in Chapter 7. It is being introduced here only because it has a current asset classification and plays a critical role in portraying the contractor's financial condition when statements are prepared on either a completed contract or percentage of completion basis.

Inventory

This current asset classification can vary in analytical significance from one type of business activity to another. For example, a general contractor subcontracting a large part of his or her work would usually show very little, if anything, in the way of building materials on hand unless some phase of the construction is performed with his or her own forces (i.e., concrete foundations, finish carpentry, etc.). An electrical, plumbing, roofing, or HVAC subcontractor, however, might be expected to show significantly more inventory than the general contractor. Cables, switch gears, pipes, shingles, sheet metal, and so on would be found in warehouses or stored on job sites. In these cases, it is important to determine what segment of the contractor's inventory is general compared to that which is purchased specifically for ongoing jobs.

General warehoused inventory may consist of various materials left over from completed jobs or those purchased in anticipation of price increases or industry-wide shortages. Future uses of these materials for a specific job may be uncertain. Since some of these may remain warehoused for months or years and eventually become obsolete, the underwriter may choose to allow only a certain percentage of general inventory as current. On the other hand, materials purchased for a specific job, referred to earlier as inventory–work in progress, will soon be incorporated into the actual job construction, become an account receivable, and eventually be converted to cash. In these cases, full asset value should be treated as current, even though the materials may be stored on or off the job site and may not actually become a part of the intended construction phase for days or weeks beyond the date of their arrival and storage. Clear distinction should be drawn between general inventory and inventory–work in progress. They have different underwriting significance in the analysis of working capital and are classified separately on the balance sheet.

While the asset value of inventory–work in progress for a contractor is usually stated by the CPA on the basis of cost, an accepted method of valuing the general inventory for any business organization would be on the basis of lower of cost or market. For example, if a plumbing contractor, in anticipation of a general price increase for copper pipe, buys large quantities for future use and the price increases as expected, the market value would be higher than cost. In that case, cost would be shown as the asset value because it is lower. If, however, the value of copper pipe decreased, the asset value would be stated on the basis of market (i.e., the lower amount being what it was actually worth).

In manufacturing type industries where large quantities of raw materials are required, the proportion or ratio of inventory to other current assets becomes much higher. As mentioned earlier, inventory in this case is broken down into three categories: raw materials, goods in process, and finished goods. As a general analytical guideline, a manufacturer's inventory should not exceed 100% of other current asset values, or be more than 50% of total current assets.

The underwriter should also be aware of the importance of inventory turnover as it relates to cost of goods sold. The formula to be used in computing inventory turnover is cost of goods sold divided by average inventory. Average is computed based on beginning and ending inventory. If, for example, annual cost of goods sold was $5,000,000 and the value of inventory was $2,500,000 ($5,000,000 divided by $2,500,000 = 2), this would mean that inventory turned over only twice a year, or every six months. The indication here might be that sales were sluggish, inventory levels were too high or obsolete, and cash flow was severely strained. On the other hand, if the average value of inventory was only $833,000, this would provide a much healthier turnover rate of six times, or every 60 days ($5,000,000 divided by $833,000 = 6).

The terms LIFO and FIFO are encountered in stating the "costs of goods sold" on the P&L and in valuing inventory. LIFO (last in, first out) would mean that costs are computed using the most recent prices paid for the raw materials used. Under the premise that costs will always continue to rise, LIFO would have the effect of producing a higher cost of goods sold and thus earnings would be stated on a more conservative basis. FIFO (first in, first out) calculates cost using the older, and usually lower, costs of materials first purchased. This method would usually result in a lower "costs of goods sold" and thus greater earnings.

In selecting either the LIFO or FIFO methods, we could have something of the credit enhancement versus a tax liability paradox. With LIFO usually treating inventory and cost of goods sold more conservatively and perhaps realistically, the prospect of encountering less tax liability through reduced earnings would be appealing, especially for a financially strong organization with adequate credit facilities. Despite this more conservative method of income recognition, some credit analysts might really feel more comfortable with the quantitative reliability of a LIFO prepared statement.

On the other side, with FIFO usually presenting a stronger financial picture, a company may be willing to incur greater tax liability if it contributed toward expansion of their credit facilities—particularly in cases where such facilities were established, in part, on the value of inventory as loan security. There could be other cases where it would be beneficial to increase a corporation's book value (net worth) prior to its sale, merger, or consolidation. FIFO may also be required where older inventories are subject to spoilage or obsolescence.

Prepaid Expenses   Prepaid expenses may arise from such prepayments as insurance premiums, advertising services, payment of quarterly income tax estimates, and interest on loans. The company is making payment in advance for benefits yet to be received, together with payment of obligations not yet due. Under GAAP these are treated as current assets on the theory that they will be used or benefit will be received within one year.

This is another area where the underwriter's treatment may differ from that of the CPA. The underwriter will classify most prepayments as noncurrent. These are not considered liquid resources available for financing the cost of construction under way, and most likely would not be canceled for purposes of obtaining a refund for the unexpired portion of the benefit provided. An exception here might be prepaid income taxes. Payment of quarterly tax estimates to IRS could be considered as an offset against accrued income tax liability, and should quarterly prepayments exceed the actual tax liability, a refund of the overpayment can be applied for. Many underwriters will, therefore, allow prepaid income taxes as current assets while disallowing the others.

All of the above lists the current assets available to retire current liabilities and has been presented in the sequence usually followed by the accounting profession. The underwriter may reclassify some of these items by treating them as noncurrent, while taking, for example, the cash surrender value of life insurance (CSVLI) usually carried under long-term or other assets and treating it as current since it can be readily redeemed for cash. Further digressions from GAAP treatment by the underwriter will be noted as current and long-term liabilities are discussed.

Fixed Assets

These assets consist of property, plant, and equipment used by the company and not intended for resale. This category includes land, buildings, machinery, equipment, furniture, automobiles, and so on, and their asset value will be declared on the basis of cost less depreciation, or "net book value."

Depreciation represents a provision for the anticipated decline in value of an asset through wear, tear, age, and obsolescence over its useful life. For example, a new dump truck costing $30,000, with an expected useful life of five years would be depreciated at a "straight line" rating of $6,000 per year until it became fully depreciated ($30,000 divided by 5 = $6,000 per year). Accelerated depreciation methods may also be used for tax accounting in which more depreciation is taken in the first years of the asset's life and less taken later. One of these methods is known as the "double declining balance" method. The notes to the financial statements should describe the various methods used for depreciating each type of fixed asset, for both book and tax purposes. Land is never depreciated and always carried at cost.

The following is how the truck and its straight line depreciation would appear on the balance sheet under fixed assets:

FIRST YEAR

Truck (cost)	$30,000	
Less accumulated depreciation	6,000	$24,000

SECOND YEAR

Truck (cost)	$30,000	
Less accumulated depreciation	12,000	$18,000

Depreciation is an expense on the P&L, either as a direct job cost or overhead expense. This has the effect of reducing profit and thus the tax liability. When a partially or fully depreciated asset is sold, the proceeds received in excess of the depreciated or "net book value" are a source of income and become taxable at ordinary income rates in the year sold.

As with inventory, the need for fixed assets will vary from one type of contractor to another. While a general contractor may own his or her building, he or she generally does not possess large quantities of rolling stock and heavy equipment. The electrical, HVAC, and plumbing contractor may need more equipment, such as trucks, generators, and so on, to operate, but as mentioned earlier, it is the grading, paving, utility, and bridge contractors that may need large and expensive quantities of heavy equipment. In the latter cases, the underwriter could be faced with an unbalanced financial condition. In this same context, if a contractor decides to lease equipment rather than buy it and the lease agreement contains a purchase option, certain accounting principles require that the leases be "capitalized," with the current and long-term lease payments treated on the balance sheet in the same manner as payments due under a note for the purchase of the equipment.

Other Assets

These assets resemble prepaid expenses except that the benefits to be derived usually extend well beyond 12 months. Other assets consist of organization cost, loan costs, research and development costs, and so on. As with depreciation, the cost is carried as an asset and written off or amortized over a specified number of years until it finally disappears from the balance sheet. These annual write-offs are also expensed on the P&L, providing the company with some tax benefit through reduced net earnings. Such assets have no physical existence and are considered "intangible assets." As such, they are eliminated entirely by the underwriter in analyzing tangible net worth. Goodwill would also fall into the category of intangible assets, representing the differences between the price of an acquired company and the related value of net assets actually acquired.

As mentioned earlier, the cash surrender value of whole life insurance policies (CSVLI) is also carried under the other asset heading. As owner of the policy, the company can either redeem the cash value accumulated since the policy's inception date or borrow against it at interest rates usually well below existing prime lending rates. Because it is readily convertible to cash, many underwriters will reclassify it as a current asset in their analysis. If it is reclassified, any policy loans should also be treated as current liabilities.

Current Liabilities

Following the same timing principle of current assets, any obligations due and payable within 12 months are classified as current liabilities. Each category will be examined as it generally appears on the balance sheet.

Notes Payable   The current maturities of long-term debt appear here (those due within 12 months from the statement date) and all balances due on short-term working capital lines. It is important for the underwriter to determine how these obligations are secured and guaranteed. If any current assets (usually receivables and/or inventory) have been pledged as collateral, the underwriter may request a letter from the lender agreeing to the exemption of receivables from the loan agreement on bonded contracts (remember the legally entrenched doctrine affording rights of "equitable subrogation" to the surety on bonded contracts, notwithstanding any UCC filings of record against these assets by a lender).

Notes payable to officers of a corporation, affiliates, or other unrelated parties may also exist. Depending on the size of these obligations, their impact on the ratio between current assets and current liabilities, the extent to which they impair working capital and the predisposition of the note holders (payees), the underwriter may require that these obligations be subordinated to the interests of the surety. In so agreeing, the note holder would execute a subordination agreement on the surety's form, agreeing to neither accept nor demand repayment without the surety's written consent. Some subordination agreements permit repayment according to the terms of the note (usually in the case of long-term notes where scheduled repayment is amortized on a monthly basis exceeding 12 months), while others allow no repayment of principal, but will permit periodic payment of accrued interest. In any form of insolvency proceedings involving liquidation of company assets, the surety's rights of recovery would take precedence over those of the subordinating note holder.

Once a note has been subordinated, the entire amount (when no principal payment is permitted) may be treated as noncurrent and carried "below the line" as a long-term liability or increase in net worth, in the underwriter's analysis. Where payment of principal is permitted, the balance due beyond 12 months may be treated as an increase in net worth. A word of caution—it is not uncommon for subordination agreements to be breached, inadvertently or otherwise. A clear verbal or written explanation by the underwriter to the parties involved as to the exact meaning and intent of the agreement at the time this instrument is executed may prevent future misunderstandings or "slipups."

Another word of caution—any notes payable to material suppliers, where normal selling terms would be on an open account basis, could signal cash flow or working capital problems and should be thoroughly investigated. This investigation may include a general trade survey conducted with other suppliers to determine the extent of the problem.

Accounts Payable   These are amounts due primarily on invoices from material suppliers, leasing companies, and any other vendor selling goods and services to the company, including invoices for operating and overhead items.

It is not uncommon for the company to prepare an aging of these accounts. This is useful in determining just how promptly the company is meeting its obligations. Such aging would be grouped into the same time categories as provided for accounts receivable aging.

Due Subcontractors   This category would usually appear on the statements of general contractors or other prime contractors normally subcontracting a good part of their work. The current billings from the subcontractors, and the retainages withheld from them on past requisitions, are usually reported here in the same proportion as that withheld from the GC by the owner. Occasionally, friction between the general and the subcontractor may occur if, for example, an owner reduces retainage withheld under the general contract from 10% to 5% at 50% completion, and these reductions are not passed along to the subcontractors. Subcontract terms should provide clarity on this point. With the specialty trades subbing very little, any amounts due their subcontractors would probably be carried in accounts payable.

Billings in Excess of Costs or Billings in Excess of Costs and Estimated Earnings (Overbillings)   This is the opposing construction accounting category to the underbilling asset, both of which are covered in Chapter 7.

Income Taxes   These are also covered later.

Accrued Expenses   These represent all other current obligations owed but not yet paid, such as salaries and bonuses accrued since the preceding pay period, interest accrued on notes payable, payroll taxes withheld from employees' paychecks, and so on.

This concludes the review of current liability categories. The total of these on the balance sheet, subtracted from current assets, gives what is commonly referred to as "working capital" or "net liquid assets," the subject of very keen interest to the underwriter. To further refine the ability of the company to quickly retire current debt, inventory and prepaid expenses are eliminated from current assets to arrive at "net quick assets." This is also referred to as the "acid test." If current liabilities exceed current assets, a *deficit working capital* is said to exist. Similarly, if total liabilities exceed total assets, but not total paid-in capital, a *retained earnings deficit* would exist. If total liabilities exceed total assets and paid-in capital, there would be a *deficit net worth* or a state of *insolvency.*

Long-Term Liabilities

Maturities of debt exceeding 12 months and loans against CSVLI would be found under this classification. Some deferred taxes and subordinated notes payable may also be shown here. While the CPA may show these two items as current liabilities, from an underwriting standpoint, some sureties will consider them long term.

Occasionally, there may be another long-term liability described as debentures or debenture bonds, so that a brief explanation of this form of obligation is appropriate. These are formal promissory notes with maturity dates extending well into the future. The notes or bonds are given to the "subscribers" or lenders as evidence of corporate debt, specifying the interest rates to be paid and the maturity dates. As with other unsecured promissory notes, they are backed only by the general credit of the company.

In some cases, these will be described as subordinated debentures, where under specific covenants in the notes, they are subordinated to general creditors in the event of liquidation.

Equity or Net Worth

There are essentially three basic forms of organizations for conducting a business. While future discussions and examples will deal with corporations, it is important to understand the other two forms before examining the corporate entity. They are the sole proprietorship and the partnership. As stated earlier, regardless of the type of business organization, equity or net worth represents the difference between total assets and total liabilities. The types of business organizations are the subject of the next chapter.

QUESTIONS—CHAPTER 5

The Financial Statement

1. Contractor A purchases a front-end loader with a useful life of seven years for $42,000. After three years, what would its book value be?

_____

_____

2. Pierce Utility Contractors, Inc. purchased 3,000 lineal feet of 24-inch ductile pipe on 11/15/97 for $10,000 in anticipation of an industry-wide shortage. They had immediate need for 2,000 lineal feet on jobs currently in progress and stored the remainder in their warehouse. The shortage did not materialize as expected and at 4/30/98 when their fiscal statement was prepared, this pipe was only worth $2.50 per lineal foot. Assuming none of the stored pipe had been used since its purchase, what amount did the CPA declare as the inventory asset value on the balance sheet, using lower of cost or market treatment?

---

3. At 12/31/97, Jergens Painting Corp.'s financial statement consisted of $3,700,000 in current assets and $2,100,000 in current liabilities. On 3/14/98, they purchased trucks, scaffolding, ladders, and so on, for $300,000, on 9/11/98 they loaned $50,000 to one of the officers, and on 6/15/98 they sold a parcel of land for $125,000. As you review this statement on August 10, and are aware of these subsequent transactions, how would you then analyze working capital?

---

4. What useful purpose does the statement of cash flows serve in analyzing a financial statement?

---

5. How would a financial statement be impacted by the write-off of an uncollectible account receivable?

---

6. Describe the difference between earned estimates and earned revenues.

---

7. Describe the accounting method most conducive to computing the cost of goods sold for taxing advantages.

---

8. If the annual revenues for a manufacturing firm were $25,000,000, and its inventory at year-end was $14,000,000, what conclusions would you draw about inventory turnover and sales?

---

9. Why do most sureties differ with regard to GAAP treatment of prepaid expenses?

_____

_____

10. Describe the surety's right of equitable subrogation to accounts receivable on bonded jobs.

_____

_____

11. Why would a surety require that a corporate officer subordinate his or her note receivable from his or her corporation?

_____

_____

12. Why would the acid test be more significant than net liquid assets in determining the true liquidity of a corporation?

_____

_____

ANSWERS—CHAPTER 5

1. $42,600 divided by 7 = $6,085.71 annual depreciation
$6,085.71 × 3 = $18,257.13 three-year depreciation
$42,600 − $18,257.13 = $24,342.87 book value

2. $10,000 divided by $3,000 lineal foot = $3.33 per lineal foot at 11/15/97 (cost)
$1,000 lineal foot × $2.50 per foot at 4/30/98 = $2,500 declared value

3. $3,700,000 Current assets
−2,100,000 Current liabilities
$1,600,000 Working capital at 12/31/97
−   300.00 3/14/98 purchase
$1,300,000
−   50,000 4/11/98 loan to officers
$1,250,000
+125,000 6/15/98 sale of land
$1,375,000 Working capital at 8/10/98

4. It illustrates the company's ability to generate positive cash flow in future accounting periods, clarifies the cash and noncash aspects of the company's investment and financing activities, and provides information about the cash receipts and cash payments during the current accounting period.

5. Reduce current assets and net earnings.

6. Earned estimates are accounts receivable, and earned revenues appear on the P&L as income.

7. LIFO.

8. Inventory turned over 1.70 times on sales, or less than twice a year. You could conclude that sales were sluggish; cash flow may be strained; some of the inventory was obsolete; and the firm's current assets were disproportionately heavy with inventory—comprising over 50% of total current assets.

9. Most prepaids will not materialize into cash during current accounting period and, with the possible exception of prepaid income taxes, do little to offset current liabilities.

10. The firmly established legal precedent provides the surety with prior right to the accounts receivable of the contractor client in default situations, notwithstanding the duly recorded assignment of those assets by a lender.

11. The enhancement of working capital and net worth.

12. With inventory and prepaid expenses eliminated, the remaining cash, cash equivalents, marketable securities, and accounts receivable used in determining *net quick* assets, are those that can more *quickly* be applied toward the retirement of current debt.

6

Types of Business Organizations

The three main legal entities that we see in the day-to-day operation of business are the proprietorship, partnership, and corporation. There are also other forms of legal structures, such as syndicates, joint stock companies, pools, joint ventures, estate ownerships, and business trusts. These are specialized and rare, however, and are eliminated from this discussion, except for the joint venture, which is becoming more common in the construction industry.

Because we as surety men and women are granting credit, it is extremely important for us to have a thorough understanding of who actually owns the business. We must know who is responsible for the operation of the business and the payment of obligations.

In all cases, whether the business is structured as a proprietorship, partnership, corporation, or joint venture, it is important for the principals to realize that business transactions must be kept separate from personal transactions. In too many cases, we find in operations—including partnerships and closely held corporations—that checks for personal items may be written from company accounts or checks may be written to "cash." Incoming checks may also be cashed, and the money spent, with no accounting.

The important thing here is to see that a qualified accountant has established an adequate set of books to secure the data necessary to prepare a financial statement that reflects the true condition of the operation, and includes all business transactions.

If we are considering an account that involves a combination of entities, although not necessarily by a joint venture, we must receive the financial statements of each as of the same date. The statement of our principal on the bond must be a fiscal year-end statement. We should also receive a fiscal year-end statement of the other entities, except when their fiscal year is different from that of our principal. In this case, we should receive the fiscal statement and, in addition, an interim statement prepared as of the same date as our principal's statement.

PROPRIETORSHIP

The single proprietorship, also known as sole proprietorship, or individual ownership, is the simplest form of business entity. With certain exceptions, the operation of the business and all policy decisions rest with one individual. This is the easiest form of business venture to start and, depending on the state in which it is operating, may require only a license to commence operations. The individual may operate under his or her own name, or may use a trade style—such as John Doe D/B/A Doe Construction Company.

The trade name may or may not be registered. Being registered means that the individual goes on public record with City, County, or State Officials, depending on local laws. He or she then may use a name for doing business, other than his or her own name, and have protection that no one else may use that name or trade style.

The individual operator, with few exceptions, pays income taxes on a calendar year basis, and the earnings, after taxes, are his or her sole property. The surety prefers to receive a financial statement on the individual proprietorship, taken from the books and records of the contracting business, and a separate statement on the assets and liabilities outside of the contracting venture. When we receive a combined financial statement, it is often difficult to follow the changes from one statement to another, particularly if the contractor is engaged in more than one enterprise. For example, a contractor may be losing money on his or her construction operation, which is the venture being bonded, yet generating profits on real estate speculation and the stock market. A combined statement would reflect an overall gain, obscuring losses in the area of greatest concern to the surety. Accordingly, growth, or lack thereof, in construction activity would be more evident from separate statements for each operation.

GENERAL PARTNERSHIP

A general partnership is a venture in which two or more individuals combine their resources to engage in a business, usually on a permanent basis. The partnership is formed after agreement by all parties as to the reason for the formation of the partnership and the nature of the business. The agreement also contains the capital contributed by each, and distribution of profits or allocation of losses. It also contains the duration of the partnership and the duties and rights of each partner. The agreement could be either oral or written, but for bonding credit purposes, we should satisfy ourselves that a written formal partnership agreement is in effect.

In a general partnership, each partner is responsible for the business acts of all. A partnership is also started quite simply, and needs only to comply with the license laws of various states. A partnership is a taxable entity, but does not of itself pay income taxes. The earnings of the partnership are taxable to each of the individual partners and their proportionate share of the profit of the partnership is included in their individual tax return, regardless of whether or not the profits were actually distributed to the partner.

The financial exhibits received on a partnership for bonding should include the financial statement of the partnership and of each of the individual partners, all prepared as of the same date. The fiscal year-end of the partnership, with few exceptions, would also be the calendar year-end, as in the proprietorship.

LIMITED PARTNERSHIP

Limited partnerships are organized under state laws and Federal regulations under the Uniform Limited Partnership Act. A written agreement must be filed with state officials and the partnership must comply with statutory requirements. A limited partnership closely parallels the forming of a corporation. A limited partnership permits the limitation of liability of one or more of the partners, but at least one general partner must be designated in addition to each of the limited partners. All limited partners must have an actual investment in the partnership by means of cash or tangible property but not services. Limited partners do not have a voice in the business operation, since this would make them general partners and fully liable for the acts of the partnership. As in the case of the general partnership, the surety should usually request the financial statements of the partnership and of each of the general and limited partners and, in most cases, the general indemnity of each partner.

THE CORPORATION

This is by far the most widely utilized vehicle for conducting a business. The corporation stands on its own as a separate legal entity. It owns all assets and owes all debt. The individual stockholders generally enjoy immunity from all corporate obligations, which is one of the features making incorporation very attractive. The corporation's capital is formed by a group of individuals (usually two or more) making their investment through the purchase of capital stock. Initially, these individuals, known as incorporators, apply to the state where they are domiciled (or any other if they choose) for a corporate charter. They decide among themselves the classes of stock, the number of shares of each class to be issued and authorized, and the par value of each. These factors become a part of the charter and may only be changed by a charter amendment, which is subject to state approval. The two most common classes of stock are common and preferred, with separate par values assigned to each. Par value can be any value agreed upon, with common stock usually being less per share than the preferred stock.

A hypothetical case has been developed to examine a rather typical capital structure. Assume this corporation was chartered in Georgia in 1957. It was authorized to issue 1,000,000 shares of common stock at $5 par value, and 10,000 shares of 5% preferred stock at $50 par value. Only 500,000 shares of common stock were initially issued, along with 1,000 shares of cumulative preferred. The day the company opened for business, stockholder's equity would have consisted of $2,500,000 cash paid in for common stock and $50,000 for preferred. The balance sheet would have appeared as follows:

ASSETS		LIABILITIES	
Cash	$2,550,000	None	
		Stockholder's Equity	
		Common stock $5.00 par value, authorized 1,000,000 shares, issued and outstanding 500,000 shares	$2,500,000
		5 percent cumulative preferred stock $50.00 par value, authorized 10,000 shares, issued and outstanding 1,000 shares	50,000
Total assets	$2,550,000	Total liabilities/equity	$2,550,000

There would still be 500,000 shares of authorized and unissued common stock and 9,000 shares of preferred stock available for future purchase by stockholders.

If, at a future date, 100,000 common shares were sold for $10 per share, or $1,000,000, $500,000 of this would be in excess of the par value ($5.00 par value × 100,000 = $500,000) and would be classified separately in the capital section under common stock as "capital in excess of par" or "paid-in capital."

As the term implies, preferred stock takes precedence over common in the payments of dividends and distribution of assets in the case of liquidation. In the preceding illustration, the cumulative preferred shareholders would be entitled to a fixed 5% of par dividend per share (or $2.50) when dividends were declared by the board of directors and these dividends would accumulate from year to year until they were declared. Dividends must be paid to these shareholders before any dividends can be paid on the common shares. There are various types of preferred stock that may affect the amount and frequency of dividend payments.

Common stockholders have no fixed limit on the amount of dividends they may receive, provided the dividends have been declared by the board of directors. The preferred dividends, however, must always be paid first. As the corporation prospers, higher dividends may be declared and, conversely, when corporate fortunes wane, the common stock dividends will probably fall also. While common shareholders take a greater risk in the years of flat earnings or operating losses, their rewards can be greater in times of prosperity.

In the corporate charter, the incorporators will decide which class of stock will be voting and nonvoting, voting stock having one vote per share. Voting control of a corporation can sometimes be a significant underwriting consideration, particularly in cases of closely held corporations (i.e., a situation where a limited number of privately owned shares of stock, and hence control, rests with elderly individuals). It is important to know the continuity plans of a corporation and to whom the stock will eventually pass.

RETAINED EARNINGS

Here is where it all comes together and everything is forced to balance. As profits from operations are retained from year to year, the growth in retained earnings is where these results are reflected. P&L profits at fiscal year-end are added to the retained earnings balance of the preceding years, while P&L losses and dividends declared are deducted. Under construction accounting, retained earnings are impacted by use of the various accounting methods selected for both tax and financial reporting purposes.

A review of retained earnings over several years may be helpful in assessing the commitment to growth by the stockholders. If, despite profitable operations, excessive salaries, bonuses, or dividends are paid each year, to the detriment of corporate growth, whether it be for tax avoidance purposes or for the personal aggrandizement of the stockholders, the underwriters must draw their own conclusions about the risk involved with such a company as opposed to one where earnings are permitted to accumulate.

In addition to retained earnings, there are other capital transactions that affect net worth, such as the infusion of additional capital through stock purchases, payment of dividends, and repurchases of outstanding stock by the corporation (treasury stock). A reconciliation of net worth, or statement of stockholders' equity, would fully describe these changes.

The combination of paid-in capital stock and retained earnings, or corporate net worth, represents the stockholder's investment. For any astute investor, return on investment (ROI) should be maximized to the fullest extent. If this ROI is stagnant for very long, it would seem prudent for the contractor to liquidate and invest his or her capital in less risky ventures, providing a higher yield with much less effort. But, on the whole, contractors in particular are usually highly optimistic and ardent risk takers. This is a part of what makes suretyship so exciting, challenging, and hopefully, rewarding.

It is important to recognize the difference between a corporate division, subsidiary, and affiliate. A division is merely an internal designation and not a separate legal entity, while a subsidiary is a separate corporate entity whose stock is wholly or partially owned by the parent corporation. An affiliate exists wherever two or more business organizations (not necessarily corporations) are owned or controlled directly or indirectly by the same individual.

THE PROFIT AND LOSS (P&L) STATEMENT OF EARNINGS OR THE INCOME STATEMENT

If the balance sheet is considered in terms of a photograph, the P&L would resemble a motion picture. It is in the P&L statement that the fruits of the contractor's efforts during the fiscal year are reflected. What is a fiscal year? It is a 12-month period of time during which the company selects to operate, then closing its books after that 12 months of operations. Some firms select their fiscal year-end at a point when operations are at their seasonal peak, thus affording the business an opportunity to present its financial condition at its best. This is ideal for credit purposes, but the IRS may like it as well, because of the greater tax liability that may result. Other firms may select a date during seasonal lows. Because of the financial and tax reporting options available to the contractor, deferment of taxable income is somewhat desensitized to seasonal peaks and valleys, although not entirely free of such sensitivity.

While the P&L covers a full 12 months of operations, any statements prepared at other times, often described as interim statements, show the operating results for the number of months elapsed since the close of the preceding fiscal year. For example, when a contractor with a November 30 FYE prepares a six-month interim statement, the heading of the P&L would show "operations for the six months ending 5/30/98," and would reflect revenues generated and costs incurred for that period only.

While interim statements are seldom audited by an independent CPA, they do organize and present a firm's capital resources and obligations in a professional accounting format, and serve as useful management tools in monitoring interim operating trends, as well as the current financial condition. They are also found revealing by credit grantors as mid-year supplements to the more fully examined and reliable year-end fiscal statements, upon which credit decisions are usually based. In some noncontract cases, with minimal financial guarantee risk, they may be accepted as the sole underwriting determinants.

The presentation of the P&Ls, as they would be stated for the four accounting methods, are found in Exhibits 8-3, 8-8, 8-10, and 8-13.

The descriptions of P&L classifications that follow depict the acceptable methods under which a business organization can recognize its revenues, allocate its costs, and arrive at the net profit after taxes. This figure is then carried forward to retained earnings on the balance sheet.

In arriving at gross profit, or gross margin, all revenues as defined below for each accounting method, arising from income generated during its normal course of business, would be reduced by the direct costs of their product, or cost of revenues. By broad definition, these direct costs for a manufacturer would be the cost of goods sold (raw materials purchased plus manufacturing labor costs and depreciation of manufacturing machinery). For a retailer or wholesaler, they would consist of the purchase price of the finished goods held for sale. In the case of a construction firm, its cost of materials, subcontractors, and unallocated indirect costs, as defined in components of the balance sheet, would constitute its costs of revenues.

The format of main and subtopics on the P&L would break down as follows.

Revenues

Revenue recognition for each of the accounting methods would be as follows:

Cash: All cash received during the accounting period.

Accrual: All billings during the accounting period.

Completed Contract: All receipts and accrued billings on completed jobs since their inception.

Percentage of Completion: All revenues from completed contracts, less those previously recognized in preceding accounting periods as uncompleted contracts, in addition to costs incurred and profits earned on uncompleted contracts, less those previously recognized in preceding accounting periods.

Cost of Revenues (Direct Costs)

These will be subtracted from the revenues:

Cash: All direct cost cash expenditures.

Accrual: All direct cost cash expenditures plus accrued accounts payable for direct costs.

Completed Contract: All direct cost cash expenditures for completed contracts only, plus corresponding accrued accounts payable for direct costs.

Percentage of Completion: All completed contract direct costs paid and accrued, less those previously recognized in preceding periods, in addition to cash expenditures and corresponding accrued accounts payable for direct costs on uncompleted contracts, less those previously recognized.

$$\text{Revenue} - \text{Direct Costs} = \text{Gross profit}$$

G&A Expenses

These will be subtracted from gross profit:

General and Administrative Expenses (G&A or Overhead): These expenses arising from general operating costs of a recurring nature, not specifically allocated as direct costs, reduce gross profit to net operating income (loss) and were discussed under "Components to the Financial Statement."

$$\text{Gross profit} - \text{G\&A expenses} = \text{Net operating profit (loss)}$$

Other Income and Expenses

These items will be added to or subtracted from the net operating profit or loss:

Other Income: Income produced from nonoperating and often nonrecurring sources appears under this heading, and would include gain on the sale of fixed assets (i.e., sold for a price in excess of their depreciated, or net book value) and interest earned on investments. This income is added to the net operating profit or loss.

Other Expenses: Interest expense and loss on the sale of assets would appear here. These expenses are subtracted from the net operating profit or loss.

Very often revealing analogies of cash flow adequacy and liquidity can be drawn from the capital available for short-term investment (interest earned) and the weight of debt service (interest expense).

Note: G&A, other income, and other expenses can be recognized under all four accounting methods.

$$\text{Net operating profit (loss)} + \text{Other income} - \text{Other expenses}$$
$$= \text{Net profit (loss) before taxes}$$

Provision for Income Taxes

This too must be subtracted from the net profit or loss:

Provision for Income Taxes: The prevailing federal and state income tax rates would be applied to whatever net profit is reported under the four accounting methods. Deferred taxes could arise from the temporary differences in depreciation, as discussed under the balance sheet, and the difference in income recognition, which is treated in Chapter 7.

$$\text{Net profit (loss) before taxes} - \text{Provision for income taxes}$$
$$= \text{Net profit (loss) after taxes}$$

QUESTIONS—CHAPTER 6

Types of Business Organizations

1. Orion, Ltd. was a limited partnership consisting of Bill Beck as general partner with Pat O'Reilly and Carl Krause paying $5,000 each for their limited partner

shares. A serious default occurred on a bonded job, requiring the surety to complete the work at a cost exceeding the contract balance by $50,000, plus $10,000 in unpaid bills. With no personal indemnity having been taken for this bond, what was the liability of each partner to the surety?

_____

_____

2. Phoenix Developers, Inc. was chartered in the state of Delaware with 50,000 shares of $5.00 par value common stock and 25,000 shares of $65.00 par value, 5% preferred stock. Initial subscriptions were for 35,000 common and 12,000 preferred shares, which was paid by the incorporators at the time the charter was granted. Because of a restrictive covenant in a subsequent loan agreement, there was no payment of dividends permitted for one year. Once the loan was fully paid and the corporation enjoyed an extremely profitable year, the Board of Directors declared that dividends of $1.50 per share would be paid to the common shareholders. How was this distribution paid?

_____

_____

3. Indicate true or false to the following assertions and provide the correct answers for the false statements.

A. The number of authorized shares of common stock times the par value per share equals total value of common stock on a financial statement.
T ____ F ____

_____

_____

B. The value of a company's common stock rose dramatically on their financial statement when an outside venture capitalist bought all of the remaining unissued shares for $5.00 over par.
T ____ F ____

_____

_____

C. The surety for a closely held corporate principal felt very comfortable with the Shareholder's Agreement between the founder and two of his or her key corporate officers. Although this agreement was not funded by life insurance, it did afford the officers the option to purchase his or her stock for $7,500,000 within 60 days from the date of his or her death, with the notes securing this payment to be held by the trustee for his or her estate.
T ____ F ____

_____

_____

D. Retained earnings are impacted entirely by the net earnings or losses of a corporation.

T _____ F _____

E. A net profit after taxes of $900,000 for a corporation with a $10,000,000 net worth would be considered a very sound return on investment (ROI).

T _____ F _____

F. The stock of a corporate division, held by the parent corporation, would be carried as an asset on the parent corporation's statement under the classification of "Investment in Subsidiary."

T _____ F _____

G. The profit recognized on an interim statement would reflect earnings generated since the date of the preceding interim statement.

T _____ F _____

H. Percentage of completion revenues would include all costs incurred and profits earned on uncompleted contracts during a fiscal year plus all amounts billed on jobs completed during the same accounting period.

T _____ F _____

ANSWERS—CHAPTER 6

1. Beck for the full $60,000. O'Reilly and Krause only suffered loss of the $5,000 investment.

2. They were first obligated to pay $39,000 to the preferred shareholders, that is, 12,000 shares × $3.25 per share = $39,000
   Then $1.50 × 35,000 = $52,500 to common stockholders for a total distribution of $91,500.

3. **A.** False. The number of issued and outstanding shares times the par value (if none were sold in excess of par).
   B. False. All payment over par value would be carried separately as "capital in excess of par."

C. False—unless the surety had satisfied itself that the two corporate officers could raise $7,500,000 in 60 days. Otherwise, the stock would revert back to the estate, leaving very uncertain future prospects for orderly ownership succession.

D. False. Payment of dividends and loss or gain on the sale of fixed assets—both impact retained earnings.

E. True. 9% net earnings after taxes is considered very sound.

F. False. A division is merely an internal designation and not a separate legal entity.

G. False. The profit recognized on an interim statement would reflect earnings generated since the preceding fiscal year-end.

H. False. Earnings recognized in preceding accounting periods would be deducted on both completed and uncompleted jobs.

7

Construction Accounting and Supplemental Financial Statement Data

So far, the balance sheet and P&L have been described with only brief references to several classifications that are unique to construction accounting. Unless these terms are fully understood and their significance appreciated, any attempt to effectively analyze the financial statement would be meaningless. While at first, explanations of these classifications may be confusing, they must be understood within the general scope of what they are intended to accomplish. They are intended to fairly state the financial condition of a construction firm despite the uncertainty over future cost estimates for its work, the corresponding recognition of income actually earned, and the most advantageous methods of reporting such income for financial and tax purposes. In the earlier example of the widget manufacturer, the cost of the product was known before it was sold. For the contractor, this cost is only an estimate.

To restate from earlier discussions, the accounting methods available are cash and accrual. Completed contract and percentage of completion methods are variations of accrual. While the cash and straight accrual methods of accounting are used by most business organizations, the completed contract or percentage of completion methods are unique to the construction industry. One method may be selected for financial reporting purposes and another for taxes. This gives rise to the deferment of tax liability into future accounting periods because of the temporary differences in income recognition between one method and another.

COMPLETED CONTRACT METHOD

Under this method, in addition to the standard items, the balance sheet also reflects all receivables and payables due and owing on both completed and uncompleted contracts, but the P&L reflects only those revenues from completed jobs. Thus the tax liability on profits earned from uncompleted work are deferred into future accounting periods.

With this method, the balance sheet indicates costs in excess of billings carried as a current asset, and billings in excess of costs as a current liability. Neither of these classifications has any effect on profits earned on the work completed through the statement date on uncompleted jobs, and merely anticipates that unbilled costs will eventually be billed and become accounts receivable, while on the liability side, the contractor is charged with billing more than costs. Under the percentage of completion method, recognition is given to earned profits on uncompleted jobs, and the completed contract overbillings can switch to the asset side when costs and earned profits exceed billings. One of the paradoxes of reporting taxable income under the completed contract method is despite the severity of losses that may be sustained on uncompleted jobs, taxes must still be paid on the profit reported on the completed jobs. When this dilemma occurs, the contractor runs into the double jolt of a lose–lose situation.

There has been strong sentiment in the U.S. Congress to repeal the completed contract method altogether for purposes of more quickly generating tax revenues on uncompleted work. The 1986 tax "simplification" reform went a long way toward full repeal, but was stopped short by opposing congressional forces. The enacted reforms still permit use of the completed contract method where total average revenues for the preceding three years are under $10,000,000. If this exception cannot be met, 90% of earned profits on uncompleted jobs become taxable (90–10 rule). Tax law, however, is in effect that does not allow the contractor to defer taxes completely until the job is completed and profit is recognized on the tax return.

If the contractor uses percentage of completion for financial statement purposes and completed contract for tax, the alternative minimum tax (AMT) requires that the company compute and pay tax at a 20% rate on the difference between taxable income per the financial statements and the tax return. There are various complex computations required to arrive at the amount of tax, but, in the long run, what AMT accomplishes is that only a small part of the tax due on uncompleted jobs is deferred. In essence, a prepayment of tax is made because, in the year the job is completed, an AMT credit is allowed for the tax previously paid; this offsets or reduces the actual tax currently due.

For political reasons, therefore, completed contract tax reporting may be on the endangered species list, since there is still strong sentiment among powerful forces "on the Hill" for total repeal.

PERCENTAGE OF COMPLETION METHOD

Here we have almost the same concept as completed contract except (1) the balance sheet carries cost *and estimated earnings* in excess of billings as a current asset, and billings in excess of cost *and estimated earnings* as a current liability, and (2) costs incurred plus earned profits on uncompleted jobs are reflected on the P&L with those from completed jobs. By inclusion of uncompleted job revenues and costs on the P&L, a stronger financial condition is usually presented. However, a greater income tax liability may also result.

Earned income or profit calculation must first be considered. It is determined by the "cost-to-cost" ratio (i.e., the percentage of costs incurred at the statement date to the total estimated costs of the job at completion). This percentage is then applied against the total revised estimated profit at completion, producing the earned profit.

Consider the following hypothetical example: A $2,000,000 job is bid with a 10% gross profit, or $200,000—leaving $1,800,000 as the total estimated cost. Six months later, $900,000 in costs have been incurred, or 50% of the total estimated costs. This is the percentage factor applied against the $200,000 total estimated gross profit at completion, thus producing $100,000 earned profit at the statement date. This earned profit is then added to the $900,000 costs incurred to arrive at total *earned revenue* of $1,000,000. Consider that billings on the job at six months are $980,000, or $20,000 less than the revenue. This difference would be expressed as cost and estimated earnings in excess of billings under current assets. On the balance sheet, it would be expressed as:

Costs ($900,000) and estimated earnings ($100,000)
    in excess of billings ($980,000) $20,000 (underbilled)

Suppose billings had been $1,050,000 with cost and earnings the same. Billings would then have been in excess of costs and estimated earnings as a current liability as follows:

Billings ($1,050,000) in excess of costs ($900,000)
    and estimated earnings ($100,000) $50,000 (overbilled)

Construction costs can fluctuate. If, in this same example, the contractor was able to accomplish a savings through "buying out" some materials at a better price and total anticipated costs were, therefore, only $1,750,000, this would leave $250,000 as total estimated profit. Costs incurred would be the same and billings at six months would be $1,200,000. Here is how this would appear:

Step 1: $900,000 (cost incurred) divided by $1,750,000 (total costs) = 51.4% earned of the total estimated profit of $250,000, or $128,500 as earned income or profit.

This would appear on the balance sheet as follows:

Step 2: Billings ($1,200,000) in excess of costs ($900,000) and estimated earnings ($128,500) = $171,500 (overbilled)

Following is this same project as it would appear on the P&L as earned revenue:

Billings	$1,200,000
Less: Overbillings	171,500
Earned revenue	$1,028,500

Percentage of completion is considered to be the method that most accurately portrays the financial condition of a contractor when the contract exceeds 12 months completion time. It is, therefore, recommended for financial reporting purposes. The completed contract method is the more popular for tax reporting. It should be pointed out that through recognizing earned profit on uncompleted jobs as an asset, as well as a part of earned revenues, the balance sheet and P&L can be grossly distorted by overstated profit estimates—more specifically estimates of costs to complete (costs incurred are known).

So far in these examples, profits have been shown. However, losses do occur, and once they become known, the entire anticipated loss should be recognized in the current accounting period or fiscal year and should be charged against revenues for that year.

DEFERRED INCOME TAXES

When one method is used for financial reporting and another for taxes, a deferred tax liability or asset may result. At present, there is much controversy within the accounting profession as to how to calculate deferred taxes. In the past, the *income statement approach* was "GAAP" and the only approach used. This is the method applied to all examples in this chapter.

With the introduction of Statement of Financial Standards (SFAS) 96, a new method was available—the *balance sheet approach*. In this method, deferred taxes are provided for by virtually all differences between financial and tax basis of assets and liabilities, as opposed to just "temporary" or timing differences. Because of the controversy and complexity of implementing this method, the profession postponed mandatory implementation until December 15, 1991.

INCOME STATEMENT APPROACH

Consider a contractor whose net income before taxes is $100,000 for financial statement purposes, but whose tax return shows only $20,000. The difference in income before and after taxes is made up of "temporary" differences (i.e., differences that will go away with the passage of time). These differences result primarily from different fixed asset depreciation methods and profit recognition methods used for book and tax purposes. To compute deferred taxes, the prevailing tax rates (i.e., 34% for example purposes) are applied to both the $100,000 percentage profit and the $20,000 cash profit. The resulting difference in tax is reported as deferred tax and is added to, or subtracted from, the prior year's deferred tax liability on the balance sheet. The $34,000 ($100,000 × 34%) tax would be the total tax expense on the financial statements.

BALANCE SHEET APPROACH

As previously stated, in the balance sheet approach to deferred taxes, virtually all differences between financial and tax basis of assets and liabilities are considered, rather than just the temporary differences. In calculating the deferred tax, liabilities are based on the higher of regular tax or alternative minimum tax. In essence, current year's rates are projected and applied to future years' differences. This projected tax is then totaled and is reflected as a liability on the balance sheet. The tax or difference for one year is shown as a current liability; the rest, which is applicable to future years, is shown as a long-term liability. The adjustment (plus or minus) needed to bring the liability balance to actual is the amount shown in the P&L as the current year's deferred tax. That means that both the P&L and balance sheet will reflect the current year's actual tax computed and due, as well as the remaining computed deferred tax related to future years.

How to treat deferred income taxes arising from temporary differences as recognition of income has been the subject of controversy among many knowledgable underwriters. Is it a current or a long-term liability? What might appear as deferred income today could disappear as officers' bonuses tomorrow. Also, there are other tax avoidance ploys that are available, so that for purposes of the examples in this text, all deferred tax liability arising from the differences in income recognition has been treated as current.

If a contractor sustains a net operating loss (NOL) on its tax reporting method, that loss can first be carried back as a reduction against taxes paid over the preceding three years and a refund applied for. This tax reporting loss can also be carried forward as a reduction to future tax reporting profits for the next 15 years. Where such tax loss carry forward exists, it should be stated in the notes to the financial statements. It may also be considered by the underwriter as an offset against whatever deferred tax liability may have been accrued on the balance sheet for financial reporting, though it has not been reflected as a reduction of the tax liability in the financial statements themselves.

SUB S CORPORATIONS

With reduction of the individual income tax rates created by the 1986 Tax Reform Act, a rash of conversions was made by many contractors from C Corporation tax filing status (where earnings are reported by the corporation) to Sub Chapter S of the Internal Revenue code. As a Sub S Corporation, all corporate net earnings are reported by the individual stockholders on their personal income tax returns. There is usually no corporate tax liability. The IRS has established limitations, however, on the maximum number of stockholders permitted (35) for Sub S status, and there can only be one class of capital stock issued.

In conversion to Sub Chapter S, the balance sheet and P&L would no longer carry any provision for current deferred taxes after the date of conversion. This would have the effect of increasing retained earnings for a healthier financial picture. On the downside, the individuals generally may have to make sizable withdrawals from the corporation in the form of higher salaries or bonuses to meet their increased personal tax liability since earnings of the corporation are now taxed to them. It is important for the underwriter to determine with some reasonable degree of accuracy how much these additional withdrawals will be. The CPA is frequently in a position to provide a reasonable estimate, particularly if the stockholder's personal tax returns are also prepared by the accounting firm. With this estimate, underwriters may make an adjustment in the analysis of working capital to provide for these additional withdrawals.

When actual distributions of all profits are made to the stockholders of a Sub S corporation at the end of the year, growth in working capital and net worth can be greatly impaired. It is, therefore, a good idea to have an agreement with the contractor to limit distributions to normal salary and bonus levels, plus only the additional amount necessary for personal tax liability.

PRO FORMA STATEMENTS

As financial forecasting is often necessary for budgeting purposes, assumption of future sales, costs, and capital requirements enable a firm to forecast operating results, cash flow, and increases in retained earnings through a given period. These assumptions can apply to the cumulative effect of monthly sales and cost projections based on historical trends, market conditions, and so on, or to a single transaction.

If, for example, a firm planned to sell a large asset for cash subsequent to the date of its latest FYE financial statement, which was encumbered by sizable debt, a pro forma statement would forecast the effects of the sale as it would retrospectively impact the financial condition existing at the statement date. Assume this asset had a book value of $300,000, corresponding debt of $150,000, and was sold for $400,000, subject to a 40% capital gains tax. The effect of this sale would be to increase working capital by $210,000, and retained earnings by $60,000.

Retained Earnings		Working Capital	
$ 300,000	Book value	$400,000	Sale
400,000	Sales price	−(150,000)	Debt
$ 100,000	Gain on sale	$250,000	
(40,000)	Tax	−(40,000)	Tax
$ 60,000	Net increase to retained earnings	$210,000	Increase

The foregoing is the method that enables an analyst to retroactively adjust certain financial statement components, based upon quantifiable developments occurring subsequent to the statement date. Of equal or greater importance is the ability to forecast future earnings and expenses from far less quantifiable assumptions.

For example, consider a contractor with a $3 million backlog at his or her latest FYE, from which it can be determined that there is $210,000 (7%) unearned anticipated gross profit to be earned during the current accounting period. For purposes of forecasting annual operating results based only on the backlog income, the estimates of G&A expenses, other income, and other expenses would have to be projected for the full year. Assume in this case that, after conferring with the contractor and his or her CPA, the underwriter determined that the estimate for G&A expenses is $400,000, other income should be $40,000 and other expenses $12,000. Here is how the forecast would appear:

Revenues	$3,000,000
Cost of revenues	2,790,000
Gross profit	210,000
G&A	(400,000)
Other income	40,000
Other expenses	(12,000)
Estimated loss	($162,000)

As these estimates were made during the first quarter of the current year and no new work had been acquired in the interim, it is obvious that the contractor will have to secure additional work that would develop gross profit of at least $162,000 before the current FYE just to break even. Assuming further that the contractor historically averaged 7% gross margins on his or her contracts, this would mean he or she would have to generate additional earned revenues of at least $2,300,000 to reach the break-even level. As contractors are not in business to break even, this would involve acquiring new contracts totalling well in excess of $2,300,000 with completion dates extending into future accounting periods.

Carrying these subjective conclusions still further, the underwriter must consider if his or her contractor client has the capacity to undertake the volume of work necessary to produce, at the very least, break-even revenues. If so, it should then be determined if the new work can be procured and completed within what could be a compressed time frame without placing an undue strain on the contractor's capital and organizational resources—bearing in mind that the closer the work is secured to the FYE, the greater the strain to produce the monthly "run-off" needed. If the contractor's bonding capacity is not adequate for him or her to undertake the volume of work needed, the underwriter is faced with the dismal prospect of his or her client suffering a loss for the year. At this point, once an estimate of the severity of the loss can be determined, the surety must make a judgment as to whether it still wishes to continue with the account and, if so, on what basis. Perhaps a scale-back in operations combined with meaningful overhead reductions could be one answer to the problem.

One of the principal lessons to be learned from these examples is to closely examine a contractor's backlog at year-end, as well as the gross margins remaining to be recognized. These should then be assessed together with the volume of new work and gross margins needed to cover anticipated expenses for the year.

Tracking individual job performance on uncompleted work, together with a review of the contractor's past gross profit trends, can provide a valuable indicator of future earnings trends through determining:

1. How close final profits have come to original bid profits;
2. The average percentage of gross profits realized; and,
3. The timeliness with which past jobs have been completed.

These trends may be followed most effectively from year-end and interim work in progress reports, together with periodic consultation with the contractor during the year. The greater the variances between bid and final profits and gross profit percentages, the less reliable the forecasts will become.

This type of forecasting is at best an inexact science, but it does establish some basis for anticipating the work program levels needed by a contractor, which may then be equated with his or her capacity to successfully undertake the volume of work required. More importantly, by constantly tracking individual job performance and profit trends, the surety establishes another early warning system for problems that may be ahead.

Before closing on this subject, it should be pointed out that, during soft construction markets in particular, revenue shortfalls resulting from the timing in securing new work are not an uncommon experience for most construction firms. The losses resulting from this problem are usually foreseen by the contractor and mitigated by his or her own initiative in reducing direct operating costs and overhead during the year. Where this happens occasionally, and there is a healthy backlog of work to be carried into the next year, there is usually little cause for alarm on the surety's part. The greater cause for alarm would be the anticipation of a year-end loss resulting from poor job performance.

REGULATORY FINANCIAL REPORTING

Many corporations and other business organizations are required to file periodic financial qualifying reports with various governmental regulatory authorities for license to conduct business in the public sector. These are usually prepared with full opinion by an independent CPA on a GAAP basis. Each of these "watchdog" bodies promulgate and monitor compliance with the financial standards it established for the pursuit of a particular line of business. In the case of construction firms, parameters are often set for a limit to which they may commit their capital and other organizational resources—translated into the single job size and aggregate work programs permitted.

A brief sampling of two such regulatory authorities would be:

SEC: This is the federal regulator of all securities trading in the United States today. It is vested with full authority to grant, suspend, and revoke the registration of any security it feels to be financially unsound for the unwary investor. To monitor the soundness of publicly traded securities, SEC mandates the filing of 10Q quarterly and 10K annual reports, which provide full disclosure of all material developments transpiring within the corporate structure of the regulated corporations from one period to the next.

Prequalification Requirements by State, County, and Local Government Agencies: Many of these agencies will require that bidders on any of their public works projects complete a prequalification form requiring, in part, disclosure of the applicant's most recent financial condition. This prequalification data bears only a vague resemblance to the far more exacting and extensive disclosures required by the SEC.

Reliance upon either of these reports as a sole underwriting source, for either contract or noncontract consideration, would be a questionable practice, but the SEC reports, in particular, provide valuable background and operating information to the surety.

As mentioned earlier, the sureties themselves are subject to regulation by the U.S. Treasury Department, which establishes the single job size they can bond on federal projects.

INDEMNITORS' FINANCIAL STATEMENTS

As the surety requires third-party indemnity to support a bonded principal, the significance of such indemnity is measured in terms of the financial strength possessed by the indemnitor. Accordingly, the indemnitor's financial statements are also required, and can range in quality and substance almost as widely as those furnished in some noncontract cases, especially those furnished by personal indemnitors. While occasionally personal statements will be prepared by an independent CPA, many are manually prepared by the indemnitors themselves, on either the surety's form or one in use by their bank, while still others may be typed in very presentable form by the principal's bookkeeper. Even when prepared by a CPA, these statements are rarely audited, so that in general, the personal statement is usually not as reliable as those required of a bonded principal.

In all cases, it is desirable for the personal statement to include schedules of the assets and liabilities stated on the balance sheet, which would break down securities, real estate, corresponding debts, and so on. While some of these statements may be accepted at face value, the surety may choose to make independent verification of some of the larger assets and notes payable and/or require additional details and explanations.

When the indemnity of a corporate subsidiary or affiliate is required, the financial data furnished is usually of better quality than that found with personal statements and is more apt to have been prepared by an independent CPA.

In both cases, the surety may be, in part, basing an underwriting decision on the liquidity and fixed asset "borrowing power" of an indemnitor and relying upon the indemnity in terms of (1) making funds available to the principal during periods of temporary cash flow interruptions; (2) the more desperate problem of responding to the principal's capital requirements as serious operating losses are being sustained, and (3) salvage for the surety who may be incurring losses in the completion of a defaulted obligation.

Note: Personal financial statements are usually prepared based on fair market value of assets and not cost. This is a great change from the regular business financial statement that GAAP requires to be prepared on historical cost.

SUMMARY

Of the Three Cs discussed in Chapter 3, the adequacy of a company's capital is the most quantifiable and verifiable through the discipline of financial statement analysis. The professionally prepared statement can tell a most revealing story about a company's financial condition, not only at the statement date, but with a reasonable degree of accuracy, how it may appear at a future date.

Conversely, there are stories it may not tell. Totally unexpected impairment of a company's assets by the stockholders through unsound investments in unrelated ventures, fraudulent conversion of assets by a dishonest employee, and litigation over large disputed receivables would be only a few examples of unanticipated developments that can seriously impact a company's balance sheet without the user's knowledge, at least until the next statement is released. Implicit in these examples could be some element of deceit if the principal continued to avail itself of a credit line established prior to these events and they were not made known to the surety. Here the "C" for the character of the company officials might come into serious question, together with their impaired financial condition.

Therefore, while a financial statement may not tell an entire story in every case, it is still the most reliable gauge for the measurement of a company's creditworthiness, but then only if the other two Cs fully support the Three-C triangle.

QUESTIONS—CHAPTER 7

Construction Accounting and Supplemental Financial Statement Data

1. Creditable Steel Erectors, Inc. converted their financial reporting method from cash to percentage of completion as of 9/30/97. Subsequently, on a $4,000,000 contract, they had incurred costs of $1,400,000 at 3/31/98 when an interim

statement was prepared. With total anticipated costs of $3,500,000 and billings of $1,500,000 through March 31, how would this job be reflected on their interim balance sheet?

2. How would this same scenario appear if the financial reporting method had been converted to completed contract?

3. In using the cash method for tax reporting on the result found in Question 1, what would deferred income be if cash collected on this job had been $1,270,000 and cash paid was $1,260,000?

4. Describe the difference between earned income and earned revenues.

5. Describe what revenues and costs of revenues would appear on the P&L for each of the accounting basis and income recognition methods.

6. What factors affect the increase or decline in net worth?

7. What two offsets to accrued tax liability might an underwriter consider?

8. Explain the difference between a C and Sub Chapter S corporation, and underwriting considerations for the latter.

9. If a contractor sustained a sizable loss, and subsequently mortgaged his or her personal residence to replenish the depleted working capital six months into

his or her fiscal year, what financial exhibit might an underwriter require in order to evaluate the effect of the loss and capital infusion?

_____

_____

10. If the financial data requested in the preceding question could not be made readily available, what stop-gap analysis might the underwriter take?

_____

_____

ANSWERS—CHAPTER 7

1. A. $4,000,000   − $3,500,000   = $500,000
         (Contract price)   (Total costs)   (Gross profit)

    B. $1,400,000 divided by $3,500,000 = 40% complete
      (Costs incurred)

    C. 40% × $500,000   = $200,000
      (Gross profit)   (Earned income)

    D. $200,000   + $1,400,000   = $1,600,000
      (Earned income)   (Costs incurred)   (Earned revenue)

    E. $1,600,000   − $1,500,000   = $100,000
      (Earned revenue)   (Billed)   (Costs/earnings in
                                              excess of billings
                                              as a current asset)

2. $1,500,000   − $1,400,000   = $100,000
  (Billings)   (Costs)   (Billings in excess
                                          of costs as a current
                                          liability)

3. Percentage earned income $200,000
  Cash basis income:
  Cash receipts $1,270,000
  Cash expended $1,260,000   <u>10,000</u>
                                $190,000   (Deferred income)

4. Earned revenues include earned income plus costs incurred.

5. Cash—Cash received as revenues and cash paid as costs of revenues.
Accrual—Total billings as revenues and total costs incurred and accrued as cost of revenues.
Completed Contract—All billings on completed contracts as revenues and all related direct costs as costs of revenues.
Percentage—All billings on completed contracts plus earned revenues on uncom-

pleted contracts as total revenues (less revenues recognized in preceding accounting periods) and all direct costs in both cases as costs of revenues (less those
previously recognized).

6. Profits or losses carried over from the P&L to retained earnings, additional
   paid-in capital from issuance of new shares of stock, dividends declared, and
   repurchase of outstanding stock by the corporation from the shareholders (treasury stock).

7. Prepaid income taxes and net operating loss carry forwards or carry backs from
   prior years' operations that had not been utilized.

8. A C corporation pays income taxes on corporate earnings. The stockholders of
   a Sub S corporation report corporate earnings on their individual tax returns.
   The underwriter should be aware of how much the corporation will have to
   distribute to the individuals to enable them to meet their personal tax liability
   (thus depleting working capital and net worth of the corporation).

9. An interim statement reflecting these changes.

10. Make pro forma adjustments to the preceding fiscal statement by (1) deducting
    the full amount of the anticipated loss; (2) adding back the contributed capital;
    and (3) adjusting income tax accruals to reflect the impact of the loss upon the
    earnings base used previously to establish the accrual.

    If an interim work-in-progress schedule were available, the profits earned
    since the last fiscal date might also be factored into these adjustments, as well
    as additional profit projections through the next fiscal date.

    The results of this exercise should serve only to provide the underwriter
    with a very general idea of how the next fiscal statement may appear, and the
    direction operations are headed in the near term. They would rarely constitute
    the basis for extending further surety credit of any significance.

8

Financial Analysis

On the pages immediately following, you are presented with excerpts from the audited fiscal year-end statement of our erstwhile subject from earlier chapters, XYZ Construction Co., Inc. While taxes are reported on the completed contract basis, these statements have been prepared on a percentage of completion basis by the very prominent firm of CPAs, Brooks, Burns, Rivers, and Lake, who are widely recognized for their expertise in the field of construction accounting.

Also included are balance sheet and P&L statements illustrating how statements prepared using the cash, straight accrual, and completed contract accounting methods would state XYZ's financial condition and earnings. All future references, however, to statement treatment will deal only with those prepared using the percentages of completion method, unless otherwise indicated. The representations on the various statements and the discussion of their content conform with standard accounting treatment as would be applied by a well-qualified CPA and will provide the reader with a good general overview of statement analysis, as well as describe how underwriters might react to what they see.

EXHIBIT 8-1 XYZ Construction Co., Inc. Balance Sheet, 12/31/97, (Percentage of Completion)

Current Assets			Current Liabilities		
Cash		2,404,900	Current portion of long-		
Marketable securities		3,616,040	term debt		15,116
Accounts receivable		4,604,902	Accounts payable		2,604,198
Earned estimates and			Due subcontractors		3,419,060
retainage		5,916,091	Accrued and withheld		
Cost and estimated earnings			payroll taxes		21,184
in excess of billings			Accrued expenses		37,889
Inventory, at cost		216,024	Billings in excess of costs		
Prepaid expenses		106,922	and estimated earnings		2,023,817
			Income taxes payable		6,514
			Deferred taxes		769,067
Total current assets		16,864,879	Total current liabilities		8,896,845
Fixed Assets					
Property and equipment			Notes payable	21,804	
Auto	21,604		Less: Current		
Truck	18,519		maturities	15,116	6,688
Tools	4,088				
	44,211				
Less:					
Accumulated					
depreciation	26,017	18,914			
			Stockholder's Equity		
			Authorized 50,000 shares		
			common stock $5.00		
			par value; issued and		
Other Assets			outstanding 35,000		
CSVLI	13,915		shares		175,000
Intangible assets	5,054	18,969	Treasury stock, at cost		(16,000)
			Retained earnings		7,839,509
			Total Stockholders Equity		7,998,509
Total Assets		16,902,042	Total Liabilities/Equity		16,902,042

EXHIBIT 8-2 Significant Factors for the Underwriter

Working Capital[a]

$16,864,879	(Current assets)
−8,896,845	(Current liabilities)
$ 7,968,034	

Net Worth

$ 175,000	(Capital stock)		Total assets	16,902,042
(16,000)	(Treasury stock)	or		
7,839,509	(Retained earnings)		Total liabilities	(8,903,533)
$7,998,509				$7,998,509

[a]Working capital stated above is on an "as declared" basis by the CPA (i.e., before the underwriter adjusts certain current assets and liabilities for analytical purposes).

**EXHIBIT 8-3   XYZ Construction Co., Inc., Statement of Earnings
for the 12 Months Ending 12/31/97**

Revenue			Percent
Completed contracts	50,081,975		
Contracts in progress	33,581,984	83,663,959	100.0%
Less cost of revenues			
Completed contracts	46,205,537		
Contracts in progress	31,235,439	77,440,976	
Gross profit		6,222,983	7.4%
General and administrative expenses			
Officers' salaries	682,460		
Other salaries	924,587		
Rent	300,000		
Utilities	36,818		
Telephones	18,609		
Legal–accounting	23,407		
Entertainment	15,689		
Travel	27,065		
Depreciation	14,603		
Bad debts	35,000		
Bonuses	850,000	2,928,238	3.5%
		3,294,745	
Other income			
Interest earned	133,608		
Gain on sale of			
fixed assets	16,090	149,698	
		3,444,443	
Other expenses			
Interest expense		3,008	
Net profit before taxes		3,441,435	4.1%
Provision for income taxes			
Current	394,160		
Deferred	844,756	1,238,916	
Net profit after taxes		2,202,519	2.6%
Retained Earnings 12/31/96		5,636,990	
Retained Earnings 12/31/97		7,839,509	

EXHIBIT 8-4   XYZ Construction Co., Inc., Completed Contracts, 12/31/97

	Contract Price	Costs	Gross Profit	%
Kelly Office Tower	$17,690,843	$16,806,300	$ 884,543	5.0%
Miles Convention Center	23,876,460	22,324,490	1,551,970	6.5%
Cross Department Store	31,006,890	27,906,201	3,100,689	10.0%
Miscellaneous	11,414,646	10,444,401	970,245	8.5%
	$83,988,839	$77,481,392	$6,507,447	7.7%
Less: Revenues Earned in Prior Period:			$2,631,009	
Revenues This Period			$3,876,438	

EXHIBIT 8-5 XYZ Construction Co., Inc., Uncompleted Jobs in Progress, 12/31/97

Job No.	Description	Contract Price[a]	Total Estimated Costs	Costs to Date	Estimated Costs to Complete	Billings to Date	Percent Complete	Total Estimated Gross Profit	Gross Profit Earned	Revenues Earned	Costs and Earnings in Excess of Billings	Billings in Excess of Costs and Earnings
76	Smith Building	26,409,817	24,561,129	11,717,509	12,843,620	12,664,902	47.7%	1,848,688	881,824	12,599,333		65,569
70	Jones Library	15,664,003	14,724,162	6,819,617	7,904,545	7,564,367	46.3%	939,841	435,146	7,254,763		309,604
80	Johnson Hotel	29,778,013	27,544,662	12,698,313	14,846,349	15,376,532	46.1%	2,233,351	1,029,575	13,727,888		1,648,644
Totals		71,851,833	66,829,953	31,235,439	35,594,514	35,605,801		5,021,880	2,346,545	33,581,984		2,023,817

[a]It is to be assumed that all of these jobs were started subsequent to 1/1/97 and that all revenues were earned during this accounting period. If any had been earned in 1996 they would have been deducted from 1997 revenues on the P&L.

Restating revenues earned:

Total billings $35,605,801
Less: Billings in excess of costs   4,370,362 (35,605,801 − 31,235,439)
                                                   $31,235,439
Plus   2,346,545 Gross profit or income earned
                                                   $33,581,984 Revenues earned (carried to P&L)

or

Costs to date $31,235,439

Plus

Gross profit   2,246,545
                       $33,581,984

Had there been any costs in excess of billings these would also have increased revenues earned.

EXHIBIT 8-6   Tax Calculations Using a 36% Factor for Federal and State

Completed contract bases net profit from tax return	$1,094,890 × 36%	Percentage net profit before taxes	$3,441,435 × 36%
Current taxes due	$ 394,160		$1,238,916
		Less current due	394,160
		Deferred taxes	$ 844,756

Arriving at balance sheet deferred tax liability using the P&L approach:

Deferred taxes 12/31/96	$920,445
Deferred taxes 12/31/97	844,756
Difference	$ 75,689
Deferred taxes 12/31/88	844,756
Less difference	75,689
	$769,067

EXHIBIT 8-7   XYZ Construction Co., Inc., Balance Sheet, 12/31/97, (Modified Cash Method)[a]

Current Assets		Current Liabilities	
Cash	$2,404,900	Current portion of long-term debt	$ 15,116
Marketable securities	3,616,040	Accrued and withheld payroll taxes	21,184
Inventory	216,024	Accrued expenses	37,889
Prepaid expenses	106,922	Income taxes payable	6,514
		Total current liabilities	$ 80,703
Total current assets	$6,343,886		
		Long-term notes payable	6,688
		Capital stock	175,000
Fixed assets	18,194	Treasury stock	(16,000)
Other assets	18,969	Retained earnings	6,134,658
Total assets	$6,381,049	Total liabilities and capital	$6,381,049

[a]Cash method does not allow accrued expenses, depreciation, or bad debts on the P&L, where *modified* cash method does.

Working Capital	
Current assets	$6,343,886
Current liabilities	80,703
	$6,263,183

Net Worth	
Stock	$ 175,000
Treasury stock	(16,000)
Retained earnings	6,134,658
	$6,293,658

EXHIBIT 8-8 XYZ Construction Co., Inc., Profit and Loss for the Year Ended 12/31/97 (Modified Cash Method)

Cash receipts	$72,640,919
Cash disbursements	68,969,781
Gross profit	$ 3,671,138
General administrative	2,928,238
	$ 742,900
Other income	149,698
	$ 892,598
Other expenses	3,008
Net profit before taxes	$ 889,590
Provision for income taxes	320,252 (36%)
Net profit after taxes	$ 569,338
Retained earnings 12/31/96	$5,565,320
Retained earnings 12/31/97	$6,134,658

Note 1: In addition to the elimination of accounts receivable and payable on the balance sheet, all entries relating to timing differences in recognition of income were also omitted (i.e., cost and earnings (C&E) billings, Billings/C&E, and deferred taxes).

Note 2: P&L revenues given for example purposes only to illustrate how cash method would be stated. It cannot be used where annual revenues exceed an average of $5,000,000.

EXHIBIT 8-9   XYZ Construction Co., Inc., Balance Sheet, 12/31/97, (Straight Accrual)

Current Assets			*Current Liabilities*		
Cash		2,404,900	Current portion of long-		
Marketable securities		3,616,040	term debt		15,116
Accounts receivable		10,520,993	Accounts payable		6,023,258
Inventory, at cost		216,024	Accrued and withheld		
Prepaid expenses		106,922	payroll taxes		21,184
			Accrued expenses		37,889
			Income taxes payable		1,967,490
Total current assets		16,864,879	Total current liabilities		8,064,937
Fixed Assets					
Property and equipment			Notes payable	21,804	
Auto	21,604		Less: Current		
Truck	18,519		maturities	15,116	6,688
Tools	4,088				
	44,211				
Less:					
Accumulated					
depreciation	26,017	18,914			
			Stockholder's Equity		
			Authorized 50,000 shares		
Other Assets			common stock $5.00		
			par value; issued and		
Cash value of life			outstanding 35,000		
insurance	13,915		shares		175,000
Intangible assets	5,054	18,969	Treasury stock, at cost		(16,000)
			Retained earnings		8,671,417
			Total stockholders equity		8,830,417
Total Assets		16,902,042	Total liabilities/equity		16,902,042

EXHIBIT 8-10   XYZ Construction Co., Inc., Statement of Earnings for the 12 Months Ending 12/31/97 (Straight Accrual)

Revenue			Percent
Completed contracts			
Contracts in progress		85,687,776	100.0%
Less cost of revenues		77,440,976	
Gross profit		8,246,800	9.6%
General and administrative expenses			
Officers' salaries	682,460		
Other salaries	924,587		
Rent	300,000		
Utilities	36,818		
Telephones	18,609		
Legal–accounting	23,407		
Entertainment	15,689		
Travel	27,065		
Depreciation	14,603		
Bad debts	35,000		
Bonuses	850,000	2,928,238	3.4%
Other income			
Interest earned	133,608		
Gain on sale of			
fixed assets	16,090	149,698	
		5,468,260	
Other expenses			
Interest expense		3,008	
Net profit before taxes		5,465,252	6.4%
Provision for income taxes		1,967,490	
Net profit after taxes		3,497,762	4.0%
Retained earnings 12/31/96		5,173,655	
Retained earnings 12/31/97		8,671,417	

EXHIBIT 8-11   Significant Factors for the Underwriter (Straight Accrual)

Working Capital[a]

$16,864,879	(Current assets)
−8,064,937	(Current liabilities)
$ 8,799,942	

Net Worth

$ 175,000	(Capital stock)		Total assets	$16,902,042
(16,000)	(Treasury stock)	or	Less:	
8,671,417	(Retained earnings)		Total liabilities	8,071,625
$8,830,417				$8,830,417

[a]Working capital stated above is on an "as declared" basis by the CPA (i.e., before the underwriter adjusts certain current assets and liabilities for analytical purposes).

EXHIBIT 8-12   XYZ Construction Co., Inc. Balance Sheet, 12/31/97, (Completed Contract Method)

Current Assets		Current Liabilities	
Cash	2,404,900	Current portion of long-term debt	15,116
Marketable securities	3,616,040	Accrued and withheld payroll taxes	21,184
Accounts receivable	4,604,902	Accrued expenses	37,889
Earned estimates		Accounts payable	2,604,198
and retainage	5,916,091	Due subcontractors	3,419,060
Costs in excess of		Billings in excess of costs	4,370,362
billings		Income taxes payable	394,160
Inventory	216,024		
Prepaid expenses	106,922		
Total current assets	16,864,879	Total current liabilities	10,861,969
		Long-term debt	6,688
		Capital stock	175,000
Fixed assets	18,194	Treasury stock	(16,000)
Other assets	18,969	Retained earnings	5,874,385
Total assets	16,902,042	Total liabilities and capital	16,902,042

Working Capital

Current assets	16,864,879
Current liabilities	10,861,969
	6,002,910

Networth

Capital stock	175,000
Treasury stock	(16,000)
Retained earnings	5,874,385
	6,033,385

EXHIBIT 8-13   XYZ Construction Co., Inc., Profit and Loss for the Year Ended 12/31/97 (Completed Contract Method)

Revenues		
Completed contracts	$50,081,975	
Less		
Cost of revenues	46,205,537	
Gross profit	$ 3,876,438	
Expenses (G&A)	2,928,238	
	$ 948,200	
Other income	149,698	
Other expenses	(3,008)	
Net profit before taxes	$ 1,094,890	
Provision for taxes	394,160	(36%)
Net profit after taxes	$ 700,730	
Retained earnings 12/31/96	$5,173,655	
Retained earnings 12/31/97	$5,874,385	

Note: Under 1986 tax reform 90% of the $2,346,545 gross profit earned on uncompleted contracts (from the uncompleted jobs in progress schedule), or $2,111,890, would also be taxable as 1997 income because revenues exceeded the $10,000,000 three-year average.

XYZ still remains an account of EZ Bonding Company, who treasures this relationship above all of its other clients. Because of XYZ's enviable earnings record over many years, the sterling character of the individual stockholders, and the impressive accumulation of earnings since its embryonic stages as a small remodeling contractor, EZ Bonding has all but "thrown the book away," in terms of how far it will extend surety credit to its favorite client. When EZ first acquired XYZ account, just prior to the Philadelphia Administration contract, it did require that the three individual stockholders and their wives personally indemnify on all bonds without limitation of liability, by way of executing a General Indemnity Agreement. Recently, the need for continuing this personal indemnification has been questioned. However, nothing as yet has been resolved on this issue. As the XYZ account has been targeted by several competitors, EZ Bonding is still giving its request for release of personal indemnity "serious thought." In addition to the many virtues of XYZ brought out earlier, let's examine some of the percentage statement features that would make this account very attractive to all surety companies:

1. There are nearly twice as many current assets as current liabilities, or a 1.90 to 1 current ratio. This is part of the healthy "balanced" condition a surety looks for. The higher the ratio, the more assets available to serve as a cushion against unforeseen contingencies that might impair the company's ability to

meet current obligations promptly. Most sureties have established minimums they will accept as a current ratio, with perhaps a 1.02 to 1 or 1.1 to 1 being too low. (See Exhibit 8-14, Summary of Financial Ratios, at the end of this chapter.)

2. Net worth of $7,998,509 compares favorably to all current and long-term debt of $8,903,533. This is another "balancing" feature sureties are keen about. In XYZ's case, there is a debt to equity ratio of almost exactly 1 to 1; that is

8,903,533	divided by	7,998,509	=	1	to	1.11
(Debt)		(Equity)		(Equity)		(Debt)

In most cases, however, you will find debt exceeding equity, and it is this disparity that the underwriter scrutinizes. For example, while it may be perfectly acceptable to have debt twice as much as equity, four of five times may be too much. Again, in this area, most sureties will establish a maximum debt to equity relationship for underwriting purposes.

3. Receivables totaling $10,520,993 theoretically turned over 7.952 times on annual revenues of $83,663,959 (revenues divided by receivables = turnover), or an average of every 46 days (365 days divided by turnover). As receivables of $5,916,091 on uncompleted jobs included slower turning retainages, the 46-day turnover would indicate very healthy collections. Payables totaling $6,023,258 turned over 12.85 times on cost of revenues of $77,440,976, or every 28 days, which would also indicate an excellent relationship with subs and suppliers.

4. Sizable cash balances are carried with surplus cash invested in short-term marketable securities. This has been, in part, accomplished through overbillings to various owners, but also indicates that receivables are being collected promptly and profits are being realized on most, if not all, of XYZ's jobs. Within reasonable limits, overbillings are healthy and common in the industry. When these exceed the total anticipated profits for the entire job, however, they might then be considered excessive. When this occurs, it is sometimes referred to as "job borrow" or "plowback" in Chapter 9, page 139.

5. Total accounts receivable and earned estimates and retainage exceed all accounts payable and due subcontractors by $4,497,735. Here is another comfortable cushion and a potential hedge against unforeseen contingencies, such as collection of the receivables in a timely manner. Incidentally, the Notes to the financial statements indicated that $35,000 had been accrued as a reserve for doubtful accounts at 12/31/96 and that this amount was written or charged off against 1997 earnings (see G&A expenses on the P&L). All current receivables are considered to be collectable.

6. On the P&L, 7.4% gross profit margins were generated, which in today's environment are quite respectable for a general contractor performing a large volume of work. G&A expenses appear to be well under control at 3.5% of total revenues—particularly where $682,460 was paid to three officers as

salaries, in addition to the $850,000 paid in bonuses. The $133,608 in interest earned certainly indicates that cash flow remained strong throughout the year, as opposed to the very minimal interest expense.

7. Of major significance is return on investment (ROI), which was a very impressive 34.3%, that is,

$2,202,519	divided by	$6,415,198*	=	34.4%
(Net after tax profit)		(12/31/96 net worth)		

Where else could you get that rate of return on your investment after payment of handsome salaries and bonuses?

Apart from these statement observations, there is much else to recommend XYZ as a highly desirable account in any surety's book. Despite their momentary lapse in wanting to bid the garbage recycling plant in Oregon, they do perform all of their work in the Northeast where they are totally familiar with the subcontract and labor markets, as well as the economic and political climate. Moreover, they stick to the type of construction they know best and decided long ago that they just didn't need to take any "flyers" in unfamiliar areas of the country.

While occasionally their bid programs (uncompleted work plus bid amounts) have come close to $100,000,000, they have never been successful in securing work where the actual uncompleted contracts amounted to more than $85,000,000 at any one time. As their corporate working capital on an "as declared" basis was $7,968,034 at 12/31/97 and net worth $7,998,509, EZ Bonding has been underwriting this account on close to a 10% basis, or "case." It is generally considered that 10% working capital and net worth to total program is quite acceptable to the underwriter, whereas 5% may not be.

This historical 10% underwriting guideline seems to have evolved from that percentage usually withheld from payment of a contractor's monthly requisitions as a cushion for the owner/ or surety remedying faulty performance and satisfying unpaid subcontractors and material suppliers. Theoretically, this retention would constitute something close to the contractor's gross profit margin, and that 90% payment should, therefore, reimburse its monthly direct job costs. The fact that a contractor "overbilled" his or her gross profit in the earlier stages of construction, rather than taking it each month on a uniform, pro rata basis, was never understood and analyzed until the mid 1960s—along with the fact that in many cases these job profits had been fully dissipated and diverted elsewhere into financing other work before the contract at hand was completed.

In today's highly complex construction environment this approach on establishing bonding parameters seems far too simplistic, but, nevertheless, the 10% working

* $175,000	Capital stock
(16,000)	Treasury stock
6,256,198	Retained earnings
$6,415,198	Net worth

capital and net worth rule to total work on hand, or backlog, still persists as a guideline, or departure point between what is considered a "light" or "strong" underwriting case. Most surety companies today seem content to underwrite on something between a 5% and 10% basis, with a mean of $7\frac{1}{2}\%$ being generally acceptable. Many sureties also establish minimum working capital and maximum debt to equity ratio's as standards for considering a new account. With all of the "state of the art" accounting treatment being applied today to the contractor's financial statement, it must be remembered that it is still only one side of the Three-C triangle.

As mentioned before, XYZ has excellent continuity plans in place by way of a buy-sell agreement, which is fully funded by officers' life insurance. Its project management team and field organization are top flight, with most of these individuals being old timers with XYZ.

In addition to having one of the finest accounting firms as its auditors and consultants, XYZ has a computer system that can track line item costs versus budget on a daily basis and produce monthly internal percentage of completion balance sheets and P&Ls for corporate management. The CPA makes frequent visits to XYZ's office to assist the bookkeeping staff.

XYZ has also retained a very prestigious law firm to handle what little litigation it does have. Most contracts are thoroughly reviewed by this firm and exceptions taken with the owner over whatever one-sided or onerous provisions may exist in the contract documents. When occasional squabbles arise over change orders, extensions of time, or subcontractor problems, the law firm also get involved. Altogether, it would be difficult to improve upon any facet of XYZ's operations.

From an underwriting standpoint, the surety will analyze XYZ's 12/31/97 statement and see how it may differ, if at all, from the "as declared" assets and liabilities stated in the audit report; that is:

Current assets (as declared)		$16,864,879
Less:	(1) Prepaid expenses	(106,922)
	(2) 25% of inventory	(54,006)
Plus:	(3) CSVLI	13,915
Adjusted current assets		$16,717,866

1. Prepaid expenses consisted of a just recently renewed property and casualty insurance policy on which the annual premium of $109,622 was paid in advance. There was also a small prepayment of interest, but no prepayment of income taxes, since the quarterly estimates paid to IRS during the year were applied to the $394,160 completed contract tax liability, which was short by only $6,514.

2. Just about all of XYZ's inventory (stated at cost on the financial statement) consisted of materials left over from completed contracts, some of which may not be needed for contracts in progress for some time into the future, if at all. If sold, the market value may be questionable so that the EZ Bonding

underwriter decided to arbitrarily reduce the declared value by 25% to allow for this uncertainty.

2. XYZ is the owner of the life insurance policies on its officers with cash value of $13,915. There were no policy loans outstanding and CSVLI could be redeemed for the full amount with 60 days notice to the carrier.

Current liabilities (as declared)	$8,896,845

There are no adjustments necessary. If the entire note payable of $21,804 had been due to an officer, and he or she had agreed to subordinate it in a manner that would not permit any repayment of principal without the surety's written consent, then there may have been cause for adjustment.

Net worth (as declared)	$7,998,509
Less: Intangible asset	(5,054)
	$7,993,455

In this case, it was found that organization costs amounted to $12,635. It was decided to amortize this cost over five years and charge it against income at the rate of $2,527 per year. As this asset is not available to XYZ in any material sense, the underwriter eliminated it from *tangible* net worth.

While the forgoing adjustments by the underwriter have negligible impact on XYZ's strong financial condition, there can be disclosures in the Notes to the financial statements that would necessitate significant adjustments. For example, should it be disclosed that a sizable portion of accounts receivable consisted of unapproved change orders, the general inclination of most underwriters would be to disallow the unapproved amounts as a current asset, at least until they were approved. In another case, a Note may reveal that, subsequent to the statement date, the contractor purchased a sizable amount of fixed assets for cash. This would have a retroactive impact on working capital (converting a current asset into a fixed asset) and the full amount of cash paid should be deducted from working capital. On the other hand, if the Note stated that a large fixed asset had been sold for cash subsequent to the statement date, this would have the effect of retroactively increasing working capital. In still another case, a Note concerning litigation might disclose that the contractor was in the process of appealing a large judgment against the company in which the trial court had imposed a legally binding obligation upon the contractor to pay a specific sum of money to the plaintiff. Since this judgment may, or may not, be reversed by the appellate court, and the timing of the appeal would be dependent upon securing a calendar date for the hearing, the full impact of the judgment must be considered by the surety. Even if the Note may have optimistic overtones about the favorable prospects for a reversal, any such legal adjudication must be charged against current year working capital.

The importance of reading the notes to the financial statement carefully cannot be understated.

EXHIBIT 8-14 Summary of Financial Ratios

Ratio	Formula	Significance
Liquidity Ratios		
Current ratio	Current assets/current liabilities	Test of debt paying ability
Acid test (quick) ratio	Cash + net receivables + marketable securities/current liabilities	Test of immediate debt paying ability
Accounts receivable turnover	Net sales/average net accounts receivable	Test of quality of accounts receivable
Number of days' sales in accounts receivable (average collection period)	Number of days in the year/accounts receivable turnover ratio	Test of quality of accounts receivable
Inventory turnover	Cost of goods sold/average inventory	Test of whether or not a sufficient volume of business is being generated relative to inventory
Total assets turnover	Net sales/average total assets	Test of whether or not volume of business generated is adequate relative to amount of capital invested in business
Tests of Equity Position and Solvency		
Equity ratio	Owner's (stockholders)/total equities	Index of long-run solvency and safety
Owner's equity to debt	Owner's (stockholder's)/debt	Index of long-run solvency and safety
Profitability Tests		
Earning power	Net operating income/operating assets	Measure of managerial effectiveness
Net income as a percent of sales	Net income/net sales	Measure of net income as a percent of sales
Net income to stockholders' equity (return on investment)	Net income/average stockholders' equity	Measure of what a given company earned for its stockholders from all sources as a percent of stockholder investment
Earnings per share (of common stock)	Net income available to common stockholders/average number of common stock outstanding	Tends to have an effect on the market price per share for public corporations.

Market Tests

The fourth category of ratios is described as Market Tests. As these relate principally to the stability of stocks in publicly traded corporations, they were not considered to be within the purview of the contract suretyship subject.

Because the accounts receivable ratio does not contemplate retentions for construction firms that may be withheld well beyond the normal selling terms for most other business organizations, they may not be particularly meaningful in analyzing a contractor's financial statement. This is where an aging of receivables, concurrent with the most recent statement, would serve to provide an underwriter with a more accurate collection analysis than the accounts receivable ratio.

The inventory turnover ratio would only be applicable to firms that were required to carry a large investment in finished and unfinished stock in trade. This ratio, therefore, would not be calculated for most construction companies unless they engaged in some type of manufacturing or fabrication operation (e.g., steel, concrete, or ductwork assembly). Companies involved in preassembled or modular units could also fall into this category.

Note: Source for ratios: *Introduction to Accounting,* 2nd edition by James Don Edwards, Roger H. Hermanson, R.F. Salmonson, and Peter R. Kensiki. Malvern, PA, American Institute, 1991.

QUESTIONS—CHAPTER 8

Problem 1

1. Explain the significance of the following ratios from an analytical standpoint:
   A. Current 1 to 3.7
   B. Debt to equity 6.7 to 1
   C. Inventory to sales 10.6
   D. Sales to receivables 11.9

_____

_____

2. If a contractor's receivables were turning over every 120 days, what two principal factors might contribute to this slowness?

_____

_____

3. If cash balances on a contractor's balance sheet at FYE were $3,000,000, and he or she was overbilled on a percentage of completion basis by $3,000,000 on several large jobs, how significant would this be to an underwriter?

_____

_____

4. If, midpoint in completing a $10,000,000 sewer line project, the contractor encountered totally unanticipated subsurface rock and water problems, for which there was no relief granted in his or her contract, and it therefore appeared his or her expected profits of $700,000 would erode into a $200,000 loss, what accounting treatment should be applied to this development?

_____

_____

5. Noble Construction Company, Inc., had been very successful, and enjoyed many years of uninterrupted prosperity. With working capital of $2,000,000, and net worth of $2,500,000, it wanted to bid a $50,000,000 job, which would have given it a $120,000,000 bid program. What was the probable reaction of the surety?

_____

_____

6. $3,500,000 owing Noble Construction Company, Inc., from a developer—out of a total of $11,800,000 in receivables due on uncompleted jobs—was 180 days past due. The developer's bank furnishing the construction financing had been taken over by FDIC as receiver because of the poor quality of their commercial

loan portfolio, and all further funding of loan commitments had been suspended. While the CPA had disclosed this development in the notes to Noble's statement, he or she had not charged off the balance due as unrecoverable, as the prospect appeared favorable that FDIC would find a buyer for the financially troubled institution, and that funding of the outstanding commitments would eventually resume. How comfortable do you feel the underwriter was with this disclosure, and what would a charge-off of the entire balance portend for Noble (and the surety)?

_____

_____

7. How would retained earnings of a corporation be affected if it had recognized earned profit on a job of $150,000 in 1994, and in the following year discovered that the job was actually going to lose $50,000 because of poor estimating?

_____

_____

8. Complete Problem #1 for Upbeat Construction and Problem #2 for General Contractors—found on the following pages.

Problem 2—Upbeat Constructors, Inc.

Prepare a 12/31/98 balance sheet and insert retained earnings, working capital, total net worth, and tangible net worth. Prepare the balance sheet using generally accepted accounting principals. Also, compute current and debt-to-equity ratios.

Capital in excess of par	$100,000
Prepaid expenses	26,000
Furniture/fixtures (net)	110,000
Cash	150,000
Taxes payable	88,000
Goodwill	14,000
Long-term debt	400,000
Preferred stock	108,000
Accounts receivable	600,000
Plant/equipment	2,025,000
Accounts payable	466,000
Inventory	85,000
Depreciation–plant/ equipment	1,000,000
Current bank debt	60,000
Common stock	250,000
Accrued expenses	43,000
Retained earnings	———
Working capital	———
Total net worth	———
Tangible net worth	———
Current ratio	———
Debt-to-equity ratio	———

Upbeat Constructors, Inc.
Balance Sheet
12/31/98

Current Assets

TOTAL CURRENT ASSETS

Fixed Assets

Other Assets

Total Assets

Current Liabilities

TOTAL CURRENT LIABILITIES

Long-Term Liabilities

TOTAL LIABILITIES

Capital Stock

Retained Earnings

Total Liabilities & Capital

Problem 3—General Contractors, Inc.

Using the attached analysis form, construct a balance sheet, arrive at retained earnings, and fill in blanks on the profit and loss statement. Follow generally accepted accounting principles.

General Contractors, Inc.
12/31/98 Balance Sheet

Interest receivable	4,300	Short-term bank notes, unsecured	55,000
Unamortized leasehold improvements	17,500	Accrued salaries	2,400
Federal income tax refund	38,100	Due against CSVLI	2,000
Accounts receivable on completed contracts	143,220	Notes payable—officer	88,400
Equipment (net)	502,400	Accounts payable—subs	33,800
Prepaid interest	2,800	Accounts payable—material	48,900
Officers' notes receivable	10,000	Accrued FICA taxes	1,600
Retainage receivable	43,800	Accrued income tax liability	83,910
Certificate of deposit	100,000	Common stock (10,000 authorized shares at 5.00 par value, 5,000 shares issued and outstanding)	25,000
Land	14,500		
Cash surrender value of life insurance (CSVLI)	3,640		
Materials, work in process (WIP)	8,900	Preferred stock (4.00 cumulative)	10,000
General inventory	47,310	Treasury stock	4,000
City of Atlanta general obligation bond	125,000	Retain earnings	———

General Contractors, Inc.
Balance Sheet
12/31/98

Current Assets

TOTAL CURRENT ASSETS

Fixed Assets

Other Assets

Total Assets

Current Liabilities

TOTAL CURRENT LIABILITIES

Long-Term Liabilities

TOTAL CURRENT LIABILITIES

Capital Stock

Retained Earnings

Total Liabilities & Capital

General Contractors, Inc.
Statement of Earnings (Profit & Loss)
for the year ended 12/31/98

Total revenues:		7,487,000
Less:	Direct job costs	6,496,014
	Gross profit	990,986
Loss:	G&A expenses (General & administrative, i.e., Overhead)	————
Net profit before taxes		————
Less:	Taxes	83,910
Net profit after taxes		————
Plus:	Retained earnings 12/31/97	622,590
Retained earnings 12/31/98		————

ANSWERS—CHAPTER 8

Problem 1

1. A. Very unbalanced deficit working capital with current debt 3.7 times current assets.

B. With debt 6.7 to equity, most analysts would consider this excessive for a construction firm—where it may not be for a manufacturer or auto dealer required to finance larger inventories.

C. 10.6 turnover would appear very healthy, with inventory turning over every 34.4 days.

D. Excellent collections with turnover rate every 30.6 days (365 days divided by 11.9).

2. Many questionable receivables, turnover over slowly or not at all, and/or sizable retained percentages being held by the owners, and a very low work program, with little cash flow being generated.

3. Little or no correlation could be made between these two figures alone, nor assumptions made that cash balances were generated entirely by overbillings. Additional excess cash could have been invested in marketable securities or fixed assets, loaned to officers, offset by underbillings on other jobs, and so on. Only by reviewing the balance sheet in its entirety and corresponding work in progress schedules could any meaningful conclusions be drawn about the severity of the overbillings.

4. Recognize the full loss in the current accounting period.

5. When considering the 1.6% working capital case to the total bid program ($2,000,000 divided by $120,000,000), the surety very likely considered this too light, and either declined the request or recommended the infusion of additional capital, finding a joint venture partner, or something similar, in order to let them favorably consider the request.

6. Extremely uncomfortable. There is no assurance that the FDIC or some potential buyer would resume funding of the outstanding loan commitments. With $2,000,000 in working capital and $2,500,000 net worth, the loss of a $3,500,000 asset would leave Noble insolvent, and unable to complete its outstanding work program of $70,000,000 ($120,000,000 bid program less the $50,000,000 bid that was declined). No responsible surety would gamble on the uncertain resumption of funding for the developer's loan, and would most likely brace itself for a very large loss.

7. Retained earnings would decrease by $200,000. The $150,000 profit recognized in 1994 would have to be backed out, in addition to recognizing the full $50,000 loss in the current year.

8. Refer to the answer sheets for Problems 1 and 2 on the following pages.

Upbeat Constructors, Inc.

Capital in excess of par	$100,000
Prepaid expenses	26,000
Furniture/fixtures (net)	110,000
Cash	150,000
Taxes payable	88,000
Goodwill	14,000
Long-term debt	400,000
Preferred stock	108,000
Accounts receivable	600,000
Plant/equipment (cost)	2,025,000
Accounts payable	466,000
Inventory	85,000
Depreciation–plant/equipment	1,000,000
Current bank debt	60,000
Common stock	250,000
Accrued expenses	43,000
Retained earnings	495,000
Working capital	204,000
Total net worth	953,000
Tangible net worth	939,000
	(elimination of goodwill)
Current ratio	1.31–1
Debt-to-equity ratio	1.11–1

Answers to Question 8, Problem 2

Upbeat Constructors, Inc.
Balance Sheet
12/31/98

Current Assets		
Cash		150,000
Accounts receivable		600,000
Inventory		85,000
Prepaid expenses		26,000
TOTAL CURRENT ASSETS		861,000
Fixed Assets		
Plant and equipment	2,025,000	
Less depreciation	1,000,000	1,025,000
Furniture & fixtures (net)		110,000
Other Assets		
Goodwill		14,000
Total Assets		2,010,000

Current Liabilities		
Current bank debt		60,000
Accounts payable		466,000
Taxes payable		88,000
Accrued expenses		43,000
TOTAL CURRENT LIABILITIES		657,000
Long-Term Liabilities		400,000
TOTAL LIABILITIES		
Capital Stock		
Preferred stock		108,000
Common stock		250,000
Capital in excess of par		100,000
Retained Earnings		495,000
Total Liabilities & Capital		2,010,000

Answers to Question 8, Problem 3

General Contractors, Inc.
Balance Sheet
12/31/98

Current Assets			Current Liabilities		
Cash equivalents		100,000	Short-term notes		55,000
Marketable securities		125,000	Notes payable–Officer		88,400
Accounts receivable		143,220	Accounts payable		48,900
Earned estimates and retainage		43,800	Due subs		33,800
Tax refund receivable		38,100	Income taxes		83,910
Notes receivables officer		10,000	Accrued salaries		2,400
Interest receivable		4,300	Accrued FICA taxes		1,600
General inventory		47,310			
Inventory WIP		8,900			
Prepaid expenses		2,800			
TOTAL CURRENT ASSETS		523,430	TOTAL CURRENT LIABILITIES		314,010
Fixed Assets			Long-Term Liabilities		
Equipment (net)		502,400	Due against CSVLI		2,000
Land		14,500			
			TOTAL LIABILITIES		
			Capital Stock		
Other Assets			Common stock		25,000
CSVLI	3,640		Preferred stock		10,000
Leasehold			Treasury stock		(4,000)
Improvements	17,500	21,140	Retained Earnings		714,460
Total Assets		1,061,470	Total Liabilities & Capital		1,061,470

General Contractors, Inc.
Statement of Earnings (Profit & Loss)
for the year ended 12/31/98

Total revenues	7,487,000
Less: Direct job costs	6,496,014
Gross profit	990,986
Loss: G&A expenses	815,206
(General & administrative)	
(Overhead)	
Net profit before taxes	175,780
Less: Taxes	83,910
Net profit after taxes	91,870
Plus: Retained earnings 12/31/97	622,590
Retained earnings 12/31/98	714,460

Step 1: Retained earnings from balance sheet at $714,460
    12/31/88

Step 2: Less: Retained earnings at 12/31/87   622,590

Step 3: Equals net profit *after* taxes at 12/31/88   91,870

Step 4: Plus tax liability   83,910

Step 5: Equals net profit before taxes   175,780

Step 6: Gross profit   990,986

Step 7: Difference between gross profit and net   815,206 = $990,986
    profit before taxes equals   −175,780
    $815,206

9

Work in Progress Analysis and the Surety's Conclusions

Once XYZ's statement had been analyzed and "fine-tuned" as to allowable working capital and net worth, and the Notes reviewed for any other unnerving disclosures that would give rise to further adjustments, there was little else to do except wait for the 6/30/98 work in progress schedule as required by EZ Bonding as a matter of routine underwriting practice. The following schedule was secured on August 1, and it appeared that no new jobs had been started since the fiscal year-end statement. This schedule of work in progress, the corresponding analysis sheet converting these figures to percentage of completion, and explanations of this latter form follow. Determine what you can about profit trends and the general direction XYZ is taking mid-year on your own before going to the EZ Bonding underwriter's conclusions and reaction to these interim financial reports on the pages immediately following the explanation of the work in progress analysis sheet.

XYZ Construction Company, Inc.
Work in Progress Schedule
6/30/98

Job No.	Description	Contract Price Including Change Orders	Estimated Costs When Bid	Total Billed to Date	Total Costs to Date	Revised Estimate of Costs to Complete
76	Smith Building	26,409,817	24,561,129	20,016,909	17,684,806	8,106,592
79	Jones Library	15,664,003	14,724,162	13,668,192	12,488,117	2,141,997
80	Johnson Hotel	29,778,013	27,544,662	21,075,402	22,646,809	8,161,673
Totals		71,851,833	66,829,953	54,760,503	52,819,732	18,410,262

XYZ Construction Company, Inc.
Analysis of Work in Progress
6/30/98

Job	1 Total Revised Costs	2 Original Estimated Gross Profit	3 Revised Gross Profit (Loss)	4 Billings in Excess of Costs	5 Costs in Excess of Billings	6 Ratio of Costs to Date to Total Costs	7 Income Earned	8 Billed Income Earned	9 Billings in Excess of Costs & Estimated Earnings	10 Earned Income Unbilled
Smith	25,791,398	1,848,688	618,419	2,332,103		68.5%	423,617	423,617	1,908,486	
Jones	14,630,114	939,841	1,033,889	1,180,075		85.4%	882,941	882,941	297,134	
Johnson	30,808,482	2,233,351	(1,030,469)		1,571,407	73.5%	(1,030,469)[a]	(1,030,469)		
Totals	71,229,994	5,021,880	621,839	3,512,178	1,571,407		276,089	276,089	2,205,620 _540,938_ 1,664,682	(See Below)[b]

[a] Once a loss becomes known the full estimated amount is charged against earnings during the current accounting period.

[b] To allow a credit for the $540,938 difference between the $1,571,407 costs already incurred (Column 5), for which no asset value will be given, and the fully accrued loss of $1,030,469 (Column 9) has been reduced to project the full future impact as this job is fully billed and the remaining costs incurred, that is,

$29,778,013	Full contract price
−21,075,402	Bill to date
$8,702,611	Remaining to be billed
−8,161,673	Costs to complete
$ 540,938	Unbilled costs to be recovered
+1,030,469	Loss
$1,571,407	Costs incurred in excess of billings (Column 5)

XYZ Construction Company, Inc.
Revised Analysis of Work in Progress
Based on 6/30/98 Profits (Losses)
Retroactive to 12/31/97

Job	1 Total Revised Costs	2 Original Estimated Gross Profit	3 Revised Gross Profit (Loss)	4 Billings in Excess of Costs	5 Costs in Excess of Billings	6 Ratio of Costs to Date to Total Costs	7 Income Earned	8 Billed Income Earned	9 Billings in Excess of Costs & Estimated Earnings	10 Earned Income Unbilled
Smith	25,791,398	1,848,688	618,419	947,393		47.7%	294,986			
Jones	14,630,114	939,841	1,033,889	744,750		46.3%	478,690			
Johnson	30,808,482	2,233,351	(1,030,469)	2,678,219		41.2%	(1,030,469)			
Totals	71,229,994	5,021,880	621,839	4,370,362		Less	(256,793)			
							(2,346,545) Profit recognized 12/31/97			
							(2,603,338) Total profit decline			

Note 1: While there would have been minor variances in percentages of completion (Column 6), the EZ underwriter used the original ones in arriving at these revisions.
Note 2: Compare income earned (Column 7) on this sheet with Gross Profit Earned Column on the 12/31/97 uncompleted jobs in progress schedule to arrive at the underwriter's profit adjustments.

BUT FIRST—A WORD ON PERCENTAGE OF COMPLETION

Before analyzing the figures on the next several pages and doing the exercise that follows, we review just what percentage of completion represents, the advantages it offers, and what gives birth to a deferred tax liability.

In the completed contract method the reader is aware that from a list giving the costs incurred and billings on each uncompleted job, those jobs with costs in excess of billings are carried as current assets and those with billings in excess of costs are classified as current liabilities. No recognition of any gross profits *earned* is reflected on jobs in progress.

In percentage of completion method, you simply continue from this point and apply the cost-to-cost ratio to the *revised* anticipated gross profit for each job (i.e., the relationship of cost already *incurred* to total anticipated *costs*). That percentage is then applied to the revised gross profit on the job to determine the gross profit or income recognized or *earned* for that job. When these earned profits are added, billings in excess of costs computed under the completed contract method may switch to the asset side and costs in excess of billings would increase as earned gross profits are added to these already underbilled jobs. Remember that when it is determined that a job will lose money, the *entire* anticipated loss is recognized during the current fiscal accounting period and charged against income earned on the other profitable jobs in that period.

Initially, the work in progress analysis sheet may seem somewhat confusing. Bear in mind that the ultimate objective is to (1) allow credit for earned income or gross profits as most *recently revised;* (2) segregate this earned income from that part that has been billed and that remaining yet to be billed, and (3) see how these figures are transported to the balance sheet and P&L.

EXPLANATION OF ANALYSIS OF WORK IN PROGRESS COLUMNS

1. This represents the total of costs incurred plus estimated costs to complete (last two columns on work in progress schedule).
2. Estimated gross profit when the job was bid.
3. The total revised costs subtracted from the total revised contract price on the work in progress schedule.
4. Simply the amount by which billings on each job exceed the cost incurred without giving any effect to earned income.
5. The opposite of 4.
6. Costs to costs ratio explained in Chapter 7.
7. The earned profit calculated by applying the costs to costs percentage to the *revised* gross profit.
8. The amount of earned profit billed. For example, if costs of $1,000,000 had been incurred with $100,000 profit earned and billings were $1,050,000—only $50,000 of the profit would have been billed—that is

$1,000,000	Costs incurred
100,000	Profit earned
$1,100,000	Costs and estimated earnings
−1,050,000	Billings
$ 50,000	Unbilled profit

To allow full credit for the unbilled profit, $50,000 would be carried under Column 10, Earned Income Unbilled.

9. The amount by which billings exceed costs and *estimated earnings.* The billings in excess of costs in Column 4 would be *reduced* by the income earned in Column 7. To arrive at costs and estimated earnings in excess of billings for the asset side, the total unbilled costs in Column 5 plus the earned income unbilled in Column 10 would be added together. Take the Smith job, for example, to illustrate how both billings in excess of costs and estimated earnings and cost and earnings in excess of billings would be reflected on the balance sheet.

Costs	$17,684,806	(from WIP Schedule)
+ Earnings	423,617	(Column 7 on analysis sheet)
	$18,108,423	
− Billings	20,016,909	(from WIP Schedule)
% Overbillings	$ 1,908,486	(Column 9 on analysis sheet is carried to the balance sheet under current liabilities as billings in excess of costs and estimated earnings)

Now assume billings in this case had only been $17,000,000. Here is how this would be carried to current assets on the balance sheet.

$18,108,423	Costs and earnings from above
17,000,000	Billings
$ 1,108,423	Costs and estimated earnings in excess of billings

In this case on the analysis sheet, we would have had:

Column 5		Column 7	Column 8	Column 10
Costs	$17,684,806	Income Earned $423,617	(No income billed)	$423,617
Billings	$17,000,000			
	$ 684,806			

By adding costs in excess of billings of $684,806 to $423,617 unbilled income you arrive at the balance sheet asset of $1,108,423. Remember, if costs exceed billings, obviously none of the earned income has been billed.

As you probably guessed, there was shock and total disbelief at EZ Bonding's home office. How did the Smith job profit drop by $1,230,269 in six months? What happened on the Johnson job where there was a reversal from a $2,233,351 estimated profit in December to a total loss projection of $1,030,469 in June? In this last job, the profit turnaround was an astonishing $3,263,820! Without having the benefit of a 6/30/98 balance sheet and P&L, the EZ Bonding officials tried to retroactively adjust the 12/31/97 statement to reflect the impact of these subsequent developments and get some handle on how the statement would have looked in December by applying the December completion percentages to the revised profit and loss estimates reported in June.

Looking back at the 12/31/97 uncompleted jobs in progress schedule, they found that $881,824 in earned income on the Smith job and $1,029,575 on the Johnson job had become part of earned revenues on the P&L and thus a part of net after-tax profit carried forward to retained earnings on the balance sheet. In the Smith case, they retroactively applied the same 47.7% earned factor at year-end to the revised total estimated profit of $618,419 and arrived at earned income of only $294,986. This was a decline of $586,838 from the $881,824 earned on the year-end jobs in progress schedule. This would have to be pulled out of the 12/31/97 working capital and retained earnings as initially analyzed.

On the Johnson job, not only did they have to back out the $1,029,575 earned profit at year end, but would have to make a further deduction for the anticipated $1,030,469 loss—for a total reduction of $2,060,044 for this job alone on the year-end statement. On the plus side, the Jones Library job had shown an improvement in its anticipated profit and would have earned $478,690 at 12/31/97 instead of the $435,146 then recognized ($1,033,889 revised gross profit × 46.3% costs-to-costs ratio at year end = $478,690). In this case, earned revenue could be retroactively increased by the difference of $43,544.

With these revised adjustments, there is obviously going to be a substantial decline in net income before tax and a corresponding reduction in tax. The $394,160 completed contract tax expense had already been paid, so that the EZ Bonding underwriter could only make an adjustment to the deferred taxes based on the reduced earnings. The $586,838 overstatement of earned income on the Smith job and the $2,060,044 on the Johnson job would both have to be treated as reductions to the $3,441,435 net income before tax at 12/31/97, while the $43,544 increase on the Jones Library job would be used as an offset. With these subtractions and the addition, the revised before-tax net profit would have been reduced to $838,097. Applying the same 36% tax rate factor to these revised net earnings, tax liability on a percentage of completion basis would be $301,715. The initial deferred tax liability of $844,756 could, therefore, be entirely eliminated, and the new deferred tax would be a reduction of $92,445 to bring the tax per financial statements to $301,715. Here's how the underwriter's work sheet would look:

Net profit before taxes 12/31/97		$3,441,435
Reduction in earnings		
Smith	($ 586,838)	
Johnson	(2,060,044)	
Jones +	43,544	(2,603,338)
Revised before tax net earnings 12/31/97		838,097
Less: Completed contract taxes paid	(394,160)	
Plus: Tax overpayment	92,445	(301,715)
Revised after-tax net earnings		536,382
Net profit originally carried to retained earnings		2,202,519
at 12/31/97		
Adjusted reduction to 12/31/97 retained earnings		(1,666,137)
12/31/97 retained earnings from statement		7,839,509
12/31/97 adjusted retained earnings		$6,173,372

As these adjustments reduced retained earnings, so would they serve as a reduction to working capital, that is:

As originally stated	$7,968,034
Less: Reduction above	1,666,137
Adjusted working capital	$6,301,897

It should be emphasized that the foregoing exercise in trying to accurately assess the general impact of the decline in earnings is inexact and totally unscientific, but it is the best that EZ Bonding could do without an interim balance sheet and P&L.

After thoroughly digesting these new figures, a meeting was called with the XYZ principals by EZ Bonding. There was understandable concern about profit trends and just how these serious reverses occurred. The surety was also more than mildly curious as to how XYZ could get so far ahead of the owner with their billings on the Smith job. In this case, XYZ had billed well over its final anticipated profit, a practice referred to earlier as "plowback" or "job borrow"; that is:

Billings/cost and earnings	$1,908,486
Total anticipated profit	618,419
Pure job borrow	$1,290,067

At the appointed hour, the XYZ people arrived for the meeting and were prepared with the information they knew EZ Bonding would want. They explained that, on the Smith job, three unbonded subcontractors had become insolvent, leaving numerous unpaid bills to their suppliers. In bringing in other subs to complete each phase, XYZ found that not only was some corrective work needed on what was already in place, but that the new prices to complete well exceeded the unpaid balances in all three of the defaulted subcontracts. XYZ had taken the precaution with one of the subs of issuing joint checks to its suppliers; however, this was the grading sub where few, if any, supplies had been needed. When questioned by EZ on why XYZ hadn't followed its usual practice of bonding subs, XYZ advised that the grader was unbondable and, because it had worked with the other two so often in the past, management hadn't felt it was necessary to pay the extra money to have them bonded (an all-too-familiar tale of woe).

On the Johnson job, there had been problems from the very beginning. XYZ's estimator had left out the cost of concrete for the underground parking decks, which XYZ was going to construct itself. The architect had refused to approve the concrete test since specifications for the Corinthian columns in front of the building had not been met. These then had to be jackhammered down and repoured, using a different type of form. Delivery of the tilt-up wall panels had been held up by the supplier because of a fire in its plant. Accordingly, final completion would be delayed by several months, with liquidated damages running at $1,000 per day. This sorrowful scenario went on and on, with the general conclusion that it had just been a "bad job."

When asked about XYZ's comfort level with the estimates of cost to complete on both jobs, the president said management had checked and rechecked each uncompleted phase with the responsible subcontractors and knew they had a good handle on all of the remaining costs. This confidence was largely predicated on the fact that everything left to do was under subcontract and that the subs would have to bear the burden of any cost escalations, even though they were all *unbonded.*

As to the Smith job's overbilling excesses, the XYZ people were quite proud of being that far ahead of the owner. They had worked hard to "front-end load" their price so that most of the profit would be billed in the first two or three requisitions. Because of the inexperienced architect assigned to this job, they had been able to overbill way beyond their wildest expectations, and the architect still hadn't caught on.

After all of this had been unfolded, the XYZ president said his firm just had to have some new work to cover the erosions of profit on the Smith and Johnson jobs and had negotiated a new contract for a 30-story office tower in Wilmington, Delaware. The estimate was $55,000,000 and the lender was going to require that the job be bonded, with its interests named as dual obligee. A written commitment from EZ Bonding was also required, confirming that it was prepared to provide the bonding needed in this case. XYZ had wisely taken the precaution of checking out the owner's financing arrangements and had secured copies of construction and permanent loan agreements to make sure the funding was adequate for both the construction and the full development costs (land acquisition, construction loan interests, architect's fee, etc.).

After having received one bombshell on top of another, EZ officials excused themselves from the meeting to commiserate among themselves in another office. What did they really have here? On one hand, they had a very fine construction firm with proven character and ability that still enjoyed a very healthy financial condition, despite the recent problems. On the other hand, they were looking at the fact that XYZ had poor operating results at 12/31/97, particularly with no new work started and not really sure of the reliability of the costs to complete estimates on the two problem jobs. EZ had seen many times in the past that, when a job became unprofitable, the initial estimates of the loss were usually understated and the actual costs escalated as the work neared completion.

Moreover, $55,000,000 was just about twice as large as any job completed by XYZ in the past. Its estimating ability may now be open to question and its new policy of not bonding subcontractors was a matter of no little concern.

Looking at the "adjusted" 12/31/97 statement figures, EZ's staff didn't really know how close they were to the real truth, but in reviewing their calculations, they made a few observations:

1. Adjusted working capital and net worth were not about $6,300,000.
2. Looking at costs to complete at 6/30/98 of $18,410,262 plus $345,750 in unearned profit ($621,839 total estimated profit less $276,089 earned) the total uncompleted work was $18,756,012. When adding the $55,000,000, therefore, to this existing work, the total projected work program or work on hand (WOH) would be $73,756,012, less whatever they had completed or "run off" since June 30.
3. On the plus side, in comparing this projected WOH to the nearly $6,300,000 in working capital and net worth, EZ Bonding would still have a healthy 8.54% underwriting case.
4. XYZ's largest job completed in the past consisted of only eight stories, but the XYZ president was not talking about 30 stories. There may be some serious questions as to the contractor's experience and ability to build so much higher.
5. EZ Bonding's maximum single job size capacity was $25,000,000, with all of its treaty reinsurers committed to their limits. Even with special acceptances from the treaties, EZ still could not approach $55,000,000 and would have to arrange for another $20,000,000 through facultative reinsurance offerings to other surety companies.

No. There were too many negative features with the recent downslide in profits and EZ Bonding just wasn't going to take the risk. On top of that, it was going to require that XYZ have a fully audited statement as of 6/30/98 before it would consider bonding anything else. It wasn't sure how accurate the retroactive adjustments to the 12/31/97 statement really were.

When the EZ officials resumed the meeting with XYZ and advised them of EZ's decision, there was complete bewilderment on the contractors' side of the table. While the XYZ officers could understand the decision on the new office building, the requirement for an audited interim statement before bonding *anything* else seemed totally inflexible. Did this mean that despite the mutual confidence and respect that had grown between surety and contractor, XYZ's still remaining sound financial condition, and the strong personal indemnity, EZ Bonding wouldn't oblige it with a bond on even a $100,000 job? EZ Bonding's answer was in the affirmative.

This XYZ scenario, as in the one on the earlier Philadelphia Administration Building bidding, does bear resemblance to the type of interplay that can and does happen in the real surety world. With this background and the treatment of statement analysis, the following pages will deal with "polishing" your analytical skills through involvement as a "front line" underwriter and problem solver. Remember as you proceed, there are seldom any "right" answers in making an underwriting decision. Serious, totally unanticipated complications can have dire effects on the best contractors, and many of these cannot be foreseen at the early stages of a job. Application of well-proven underwriting principles, common sense, and a little luck are the ingredients needed to remain a survivor in the surety game. Most, if not all, of the "gurus" in the surety industry have had contract losses at one time. The object is to keep those losses as few and far between as possible.

Following are seven analytical exercises, the last three of which consist of case studies that treat other controversial bonding scenarios. They are presented in a challenging problem solving/solution format, where it would be helpful to the readers to arrive at their thoughts on the surety's decision before referring to the solution pages.

In the case of Problem 6, note the similarity between the commercial bank loan officer's approach to an unsecured revolving line of credit request, and that of the bond underwriter's position in extending unsecured surety credit.

PROBLEM 1

Systematic Construction, Inc.

You are furnished with a 4/30/98 work in progress schedule and requested to complete the attached analysis sheet to arrive at percentage of completion, under- and overbillings.

Systematic Construction, Inc.
Work in Progress Schedule
4/30/98

Job Description	Owner	Contract Price Including Change Orders	Estimated Cost When Bid	Total Billed to Date Including Retainage	Total Costs to Date	Revised Estimated Costs to Complete
Kinder Care	Mother Goose	16,849,002	15,164,101	7,466,617	5,904,088	10,600,922
Joe's Bar	Joe	26,498,313	23,848,481	19,488,804	19,504,920	4,412,064
Petals Boutique	Dandy Lion	4,899,407	4,409,466	2,733,918	1,648,891	3,217,631
Totals		48,246,722	43,422,048	29,689,339	27,057,899	18,230,637

Completed Contracts Schedule
Completed Since Last Fiscal Year-End

Job Description	Owner	Final Contract Price	Total Cost	Gross Profit
Totals				

Systematic Construction, Inc.
Analysis of Work in Progress
4/30/98

1 Total Revised Costs	2 Original Estimated Gross Profit	3 Revised Gross Profit (Loss)	4 Billings in Excess of Costs	5 Costs in Excess of Billings	6 Ratio of Costs to Date to Total Costs	7 Income Earned	8 Billed Income Earned	9 Billings in Excess of Costs & Estimated Earnings	10 Earned Income Unbilled
Totals									

For transference to the balance sheet complete the blanks below:

Billings in excess of costs and estimated earnings _____

Costs and estimated earnings in excess of billings _____

Answer to Problem 1

Systematic Construction, Inc.
Analysis of Work in Progress
4/30/98

	1 Total Revised Costs	2 Original Estimated Gross Profit	3 Revised Gross Profit (Loss)	4 Billings in Excess of Costs	5 Costs in Excess of Billings	6 Ratio of Costs to Date to Total Costs	7 Income Earned	8 Billed Income Earned	9 Billings in Excess of Costs & Estimated Earnings	10 Earned Income Unbilled
	16,505,010	1,684,901	343,992	1,562,529		35.8%	123,149	123,149	1,439,380	
	23,917,004	2,649,832	2,581,309		16,116	81.5%	2,103,767			2,103,767
	4,866,522	489,941	32,885	1,085,027		33.8%	11,115	11,115	1,073,912	
Totals	45,288,536	4,824,674	2,958,186	2,647,556	16,116		2,238,031	134,264	2,513,292	2,103,767

For transference to the balance sheet complete the blanks below:

Billings in excess of costs and estimated earnings   2,513,292   (Column 9)

Costs and estimated earnings in excess of billings   2,119,883   (Columns 5 & 10)

PROBLEM 2

Intrepid Contractors, Inc.

In this case you are requested to complete the balance sheet from the list of various classifications in a manner that would conform with standard accounting practices. This will, in part, require that you determine whatever percentage of completion, under-, or overbillings existed at the statement date.

Please then refer to the last sheet of this problem and list what adjustments you would make to arrive at working capital and net worth for underwriting purposes. Also, list what questions you would ask to determine the true asset value of certain classifications.

Intrepid Contractors, Inc. (GCs)
3/31/98 FYE Statement Percentage Basis

Prepaid income tax	$ 9,500
Accounts payable	38,000
Additional paid-in capital	100,000
Notes payable officer	18,700
Miscellaneous stocks	40,000
Retainage rec.	135,000
Goodwill	4,000
Notes pay bank (short-term)	50,000
Equipment (cost)	125,000
Materials	16,500
Land	70,000
Accrued FICA taxes	3,100
Depreciation (equipment)	100,000
Capital stock (common)	25,000
Note rec.	88,000
Due subs	31,000
Accounts receivable	211,000
Cash in bank	67,000
Deposits (utilities)	1,400
Allowance for bad debt	24,000
CDs	100,000
Income taxes payable	22,000
Deferred income taxes payable	49,000*
Earned surplus (retained earnings)	_____
Costs and estimated earning/billings	_____
Billings/costs and estimated earnings	_____

*Due to difference in depreciation methods, that is:

- Straight line for financial reporting.
- Double declining balance method for tax reporting (accelerated). Treat as noncurrent.

Intrepid Contractors, Inc.
Work in Progress Schedule
3/31/98

Job Description	Owner	Contract Price Including Change Orders	Estimated Cost When Bid	Total Billed to Date Including Retainage	Total Costs to Date	Revised Estimated Costs to Complete
Office Building	USA	3,935,000	3,550,000	2,840,000	2,100,000	1,275,000
U of Georgia	Dormitory	2,740,000	2,650,000	2,680,000	2,300,000	470,000
Totals		6,675,000	6,200,000	5,520,000	4,400,000	1,745,000

Completed Contracts Schedule
Completed Since Last Fiscal Year-End

Job Description	Owner	Final Contract Price	Total Cost	Gross Profit
Totals				

Intrepid Contractors, Inc.
Analysis of Work in Progress
3/31/98

	1 Total Revised Costs	2 Original Estimated Gross Profit	3 Revised Gross Profit (Loss)	4 Billings in Excess of Costs	5 Costs in Excess of Billings	6 Ratio of Costs to Date to Total Costs	7 Income Earned	8 Billed Income Earned	9 Billings in Excess of Costs & Estimated Earnings	10 Earned Income Unbilled
Totals										

Intrepid Contractors, Inc.
Balance Sheet
3/31/98

Current Assets

TOTAL CURRENT ASSETS

Fixed Assets

Other Assets

Total Assets

Current Liabilities

TOTAL CURRENT LIABILITIES

Long-Term Liabilities

TOTAL LIABILITIES

Capital Stock

Retained Earnings

Total Liabilities & Capital

*Ignore second column

Adjustments to working capital and tangible net worth. Please indicate reductions in (   ) and list below what assets and liabilities you would question.

Current assets as declared on   _____
   balance sheet
       + or (−)

                                        _____
                                        _____
                                        _____
                                        _____
                                        Adjusted current assets   _____

Current liabilities as declared   _____
   on balance sheet
       + or (−)

                                        _____
                                        _____
                                        _____
                                        _____
                                        Adjusted current liabilities   _____

Net worth as declared   _____
       + or (−)

                                        _____
                                        _____
                                        _____
                                        _____
                                        Adjusted net worth   _____

Questions:

Answers to Problem 2

Intrepid Contractors, Inc. (GCs)
3/31/98 FYE Statement Percentage Basis

Prepaid income tax	$ 9,500
Accounts payable	38,000
Additional paid-in capital	100,000
Notes payable officer	18,700
Miscellaneous stocks	40,000
Retainage rec.	135,000
Goodwill	4,000
Notes pay bank (short-term)	50,000
Equipment (cost)	125,000
Materials	16,500
Land	70,000
Accrued FICA taxes	3,100
Depreciation (equipment)	100,000
Capital stock (common)	25,000
Note rec.	88,000
Due subs	31,000
Accounts receivable	211,000
Cash in bank	67,000
Deposits (utilities)	1,400
Allowance for bad debt	24,000
CDs	100,000
Income taxes payable	22,000
Deferred income taxes payable	49,000*
Earned surplus (retained earnings)	(395,080)
Costs and estimated earning/billings	——
Billings/costs and estimated earnings	(801,680)

*Due to difference in depreciation methods, that is:

• Straight line for financial reporting.

• Double declining balance method for tax reporting (accelerated).

Intrepid Contractors, Inc.
Analysis of Work in Progress
3/31/98

1 Total Revised Costs	2 Original Estimated Gross Profit	3 Revised Gross Profit (Loss)	4 Billings in Excess of Costs	5 Costs in Excess of Billings	6 Ratio of Costs to Date to Total Costs	7 Income Earned	8 Billed Income Earned	9 Billings in Excess of Costs & Estimated Earnings	10 Earned Income Unbilled
3,375,000	385,000	560,000	740,000		62.2%	384,320	348,320	391,680	
2,770,000	90,000	(30,000)	380,000			(30,000)	(30,000)	410,000[b]	
Totals 6,145,000	475,000	530,000	1,120,000[a]			318,320	318,320	801,680	

[a]To reconcile $1,120,000 Billings in Excess of Costs you would add the following:

(USA) Billings in excess of costs & earnings $ 391,680 (Unearned)
(U of GA) Billings in excess of costs & earnings 380,000 (Unearned)
Loss of U of GA Dormitory 30,000 (Loss)
Income earned 318,320
_____
$1,120,000

[b]The $30,000 loss would be added to the pure overbillings of $380,000. As earned income would reduce this liability in conversion to percentage of completion, so would a loss add, to it.

152

Intrepid Contractors, Inc.
Balance Sheet
12/31/98

Current Assets			Current Liabilities		
Cash		67,000	Notes payable—Bank		50,000
Cash Equivalents		100,000	Notes payable—Officer		18,700
Marketable Securities		40,000	Accounts payable		38,000
Accounts Receivable Net of			Due subs		31,000
Allowance for Bad Debt		187,000	Billings in excess of costs and estimated earnings		801,680
Earned Estimates and Retainage		135,000	Income taxes—Current		22,000
Other Accounts and Notes Receivable		88,000	Income taxes—Deferred		49,000
Prepaid Income Taxes		9,500	Accrued liabilities		3,100
Inventory		16,500			
Prepaid Expenses		1,400			
			TOTAL CURRENT LIABILITIES		1,013,480
	TOTAL CURRENT ASSETS	644,400			
			Long-Term Liabilities		
Fixed Assets					
Plant and Equipment	125,000				
Less:					
Depreciation	100,000	25,000			
Land		70,000			
				TOTAL LIABILITIES	1,013,480
			Capital Stock		
			Common		25,000
Other Assets			Additional Paid-in Capital		100,000
Goodwill		4,000			
			Retained Earnings		(395,080)
Total Assets		743,400	Total Liabilities & Capital		743,400

153

Adjustments to working capital and tangible net worth. Please indicate reductions in ( ) and list below what assets and liabilities you would question.

Current assets as declared on balance sheet + or (−)		$644,400
	Miscellaneous stocks	(40,000)
	Notes receivable	(88,000)
	Inventory	(8,250)
	Prepaid expenses	(1,400)
	Adjusted current assets	506,750
Current liabilities as declared on balance sheet + or (−)		1,013,480
	Deferred taxes	(49,000)
	Adjusted current liabilities	964,480
Net worth as declared + or (−)		(270,000)
	Goodwill	(4,000)
	Adjusted net worth	(274,080)

Questions:	Answers:
• What do miscellaneous stocks consists of?	High risk unlisted securities (disallow)
• From whom is 88,000 note receivable due? Is it secured? Has it since been paid?	John Doe No No (disallow)
• Is all of inventory readily available for contracts in progress?	No. General warehouse inventory only. Allow 50%

PROBLEM 3

Honest Contractors, Inc.
12/31/98 Fiscal Statement on Percentage of Completion Method

Construct a balance sheet on the attached form both on a percentage of completion and modified cash basis, as you would allow them in *an underwriting analysis*. Assume no income tax liability in both cases. Then fill in the blanks below for the percentage of completion statement.

Capital stock (preferred)	$10,000	
Bank overdraft (treat as current liability)	(17,000)	
Cost and estimated earnings in excess of billings	_____	
Accounts receivable	148,000	
CSVLI	10,000	
Goodwill	8,500	
Inventory (for jobs in progress)	115,000	
Tax refund due	4,400	
Equipment net of depreciation	122,000	
Georgia Power Bonds	7,000	
Prepaid insurance	1,200	
Notes receivable—Officer	2,000	
Land & buildings	100,000	
Billings in excess of costs and estimated earnings	_____	
Note payable—Bank	25,000	(due on demand)
Note payable—Real estate	5,000	(due after 12 mo)
Note payable—Equipment	42,000	(due after 12 mo)
Debentures payable	50,000	(due 2011)
Accrued miscellaneous taxes	5,000	
Accrued bonuses	2,000	
Accounts payable	32,000	
Retained earnings (% method)	_____	

Honest Contractors, Inc.
Work in Progress Schedule
12/31/98

Job Description	Owner	Contract Price Including Change Orders	Estimated Cost When Bid	Total Billed to Date Including Retainage	Total Costs to Date	Revised Estimated Costs to Complete
AHA Building	City of Atlanta	2,804,900	2,491,900	2,565,119	1,817,404	874,396
I Q High School	State of Georgia	1,796,309	1,522,420	620,919	583,441	899,168
Totals		4,601,209	4,014,320	3,186,038	2,400,845	1,773,564

COMPLETED CONTRACTS SCHEDULE
COMPLETED SINCE LAST FISCAL YEAR-END

Job Description	Owner	Final Contract Price	Total Cost	Gross Profit
Totals				

Honest Contractors, Inc.
Analysis of Work in Progress
12/31/98

1 Total Revised Costs	2 Original Estimated Gross Profit	3 Revised Gross Profit (Loss)	4 Billings in Excess of Costs	5 Costs in Excess of Billings	6 Ratio of Costs to Date to Total Costs	7 Income Earned	8 Billed Income Earned	9 Billings in Excess of Costs & Estimated Earnings	10 Earned Income Unbilled
Totals									

Honest Contractors, Inc.
Balance Sheet
12/21/98

Current Assets

%   Modified Cash

TOTAL CURRENT ASSETS

Fixed Assets

Other Assets

Total Assets

Current Liabilities

%   Modified Cash

TOTAL CURRENT LIABILITIES

Long-Term Liabilities

TOTAL LIABILITIES

Capital Stock

Retained Earnings

Total Liabilities & Capital

Answers to Problem 3

HONEST CONTRACTORS, INC.

12/31/98 FISCAL STATEMENT ON PERCENTAGE OF COMPLETION METHOD

Capital stock (preferred)	$10,000	
Bank overdraft (treat as current liability)	(17,000)	
Cost and estimated earnings in excess of billings	85,806	
Accounts receivable	148,000	
CVLI	10,000	
Goodwill	8,500	
Inventory (for jobs in progress)	115,000	
Tax refund due	4,400	
Equipment net of depreciation	122,000	
Georgia Power Bonds	7,000	
Prepaid insurance	1,200	
Notes receivable—Officer	2,000	
Land & buildings	100,000	
Billings in excess of costs and estimated earnings	671,373	
Note payable—Bank	25,000	(due on demand)
Note payable—Real estate	5,000	(due after 12 mo)
Note payable—Equipment	42,000	(due after 12 mo)
Debentures payable	50,000	(due 2011)
Accrued miscellaneous taxes	5,000	
Accrued bonuses	2,000	
Accounts payable	32,000	
Retained earnings (% method)	263,967)	

Honest Contractors, Inc.
Analysis of Work in Progress
12/31/98

1 Total Revised Costs	2 Original Estimated Gross Profit	3 Revised Gross Profit (Loss)	4 Billings in Excess of Costs	5 Costs in Excess of Billings	6 Ratio of Costs to Date to Total Costs	7 Income Earned	8 Billed Income Earned	9 Billings in Excess of Costs & Estimated Earnings	10 Earned Income Unbilled
2,691,800	313,000	113,100	747,715		67.5%	76,342	76,342	671,373	
1,482,609	273,889	313,700	37,478		39.3%	123,284	37,478		85,806
Totals									
4,174,409	586,889	426,800	785,193			199,626	113,820	671,373	85,806

Honest Contractors, Inc.
Balance Sheet
12/21/98

	%	Modified Cash
Current Assets		
Marketable Securities	7,000	7,000
CVLI	10,000	
Tax refund	4,400	
Accounts receivable	148,000	
Cost & estimated earnings		
In excess of billings	85,806	
Inventory	115,000	115,000
TOTAL CURRENT ASSETS	370,206	122,000
Fixed Assets		
Equipment & plant (net)	122,000	122,000
Land & buildings	100,000	100,000
Total fixed assets	222,000	222,000
Other Assets		
Goodwill		
Prepaid insurance	1,200	
Due from officer	2,000	
Total Assets	595,406	344,000

	%	Modified Cash
Current Liabilities		
Bank overdraft	17,000	17,000
Note payable—bank	25,000	25,000
Accounts payable	32,000	
Accrued bonuses	2,000	
Billings in excess of costs and estimated earnings	671,373	
Accrued misc. taxes	5,000	
TOTAL CURRENT LIABILITIES	752,373	42,000
Long-Term Liabilities		
Notes payable—RE	5,000	5,000
Notes payable—Equip.	42,000	42,000
Debentures payable	50,000	50,000
TOTAL LIABILITIES	849,373	139,000
Capital Stock		
Preferred	10,000	10,000
Retained Earnings	(263,967)	195,000
Total Liabilities & Capital	595,406	344,000

PROBLEM 4

Imperturbable Constructors, Inc.

This statement is to be analyzed for underwriting purposes.

1. Determine unknown under-/overbilling factors on the analysis form and construct a balance sheet.
2. Fill in blanks on data sheet.
3. What was contractor's WOH at 4/30/98?
4. What is the current ratio?
5. What is the debt to equity ratio?
6. Based on outstanding WOH would you say that the underwriting basis at 4/30/98 was (check one):

Exceptionally strong	_____
Strong	_____
Average	_____
Light	_____
Inadequate	_____

7. Would the contractor be eligible for additional bonding consideration at 4/30/98? How much aggregate?
8. What other information would you want to consider this account?

Data Sheet
Imperturbable Constructors, Inc.
4/30/98 FYE Statement

Capital stock (common)	$150,000
Reserve—Doubtful accounts receivable	23,000
Accrued miscellaneous taxes	4,200
Georgia Power Bonds	165,000
Equipment notes due in one year	37,500
CSVLI	10,300
Equipment (book value)	93,700
Retainage receivable	44,000
Accounts receivable	187,300
Inventory—Work in progress	13,800
Cash in bank	111,414
Accrued bonuses and salaries	5,300
Accounts payable	74,900
Accrued income taxes	22,000
Equipment notes—Due beyond one year	41,500
Bank note payable (short-term)	100,000
Deferred income taxes	10,000 (from depreciation)
Notes payable—Officers	26,500 (subordinated)
Prepaid income taxes	4,000
Land	137,500
Unamortized leasehold improvements	55,000
Accounts receivable—Employees	1,500
Additional paid-in capital	100,000
Cost and estimated earnings over billings	——
Billings over costs and estimated earnings	——
Retained earnings	——

Imperturbable Constructors, Inc.
Work in progress Schedule
4/30/98

Job Description	Owner	Contract Price Including Change Orders	Estimated Cost When Bid	Total Billed to Date Including Retainage	Total Costs to Date	Revised Estimated Costs to Complete
Archives	USA	6,250,110	5,750,110	4,835,400	4,600,500	980,400
Zoo Addition	City of Baltimore	3,756,000	3,431,000	1,614,450	1,302,010	2,189,990
New Assembly Plant	General Motors	2,504,600	2,194,600		200,000	1,994,600
Totals		12,510,710	11,375,710	6,449,850	6,102,510	5,164,990

Completed Contracts Schedule
Completed Since Last FYE

Job Description	Owner	Final Contract Price	Total Cost	Gross Profit
Totals				

Imperturbable Constructors, Inc.
Analysis of Work in Progress
4/30/98

1 Total Revised Costs	2 Original Estimated Gross Profit	3 Revised Gross Profit (Loss)	4 Billings in Excess of Costs	5 Costs in Excess of Billings	6 Ratio of Costs to Date to Total Costs	7 Income Earned	8 Billed Income Earned	9 Billings in Excess of Costs & Estimated Earnings	10 Earned Income Unbilled
Totals									

Imperturbable Constructors, Inc.
Balance Sheet
4/30/98

Current Assets

TOTAL CURRENT ASSETS

Fixed Assets

Other Assets

Total Assets

*Ignore second column

Current Liabilities

TOTAL CURRENT LIABILITIES

Long-Term Liabilities

TOTAL LIABILITIES

Capital Stock

Retained Earnings

Total Liabilities & Capital

Answers to Problem 4

Imperturbable Constructors, Inc.

This statement is to be analyzed for underwriting.

1. Determine unknown under-/overbilling factors on the analysis form and construct a balance sheet.
2. Fill in blanks on data sheet.
3. What was contractor's WOH at 4/30/98?

Costs to complete	$5,164,990
Unearned profit	565,363
Total	$5,730,353

4. What is the current ratio?   2.30:1
5. What is the debt to equity ratio?   1:1.40
6. Based on outstanding WOH would you say that the underwriting basis at 4/30/98 was (check one):

Exceptionally strong	_____
Strong	_____X_____
Average	_____
Light	_____
Inadequate	_____

7. Would the contractor be eligible for additional bonding consideration at 4/30/98? How much aggregate?   Yes. Up to $8,000,000 or more with everything being favorable.
8. What other information would you want to consider with this account?

Performance record

Resumés of officers and key employees

Personal statement of officers and their indemnity

Bank credit

Trade reports

Last two years of corporate statements

Reference checks with architects, owners, subs, and suppliers

Completed profile form

Key man life insurance

Buy/sell agreement

Most recent work in progress report and/or interim statement

Imperturbable Constructors, Inc.
Analysis of Work in Progress
4/30/98

1 Total Revised Costs	2 Original Estimated Gross Profit	3 Revised Gross Profit (Loss)	4 Billings in Excess of Costs	5 Costs in Excess of Billings	6 Ratio of Costs to Date to Total Costs	7 Income Earned	8 Billed Income Earned	9 Billings in Excess of Costs & Estimated Earnings	10 Earned Income Unbilled
5,580,900	500,000	669,210	234,900		82.4%	551,429	234,900		316,529
3,492,000	325,000	264,000	312,440		37.2%	98,208	98,208	214,232	
2,194,600	310,000	310,000		200,000	9.1%	28,210			28,210
Totals									
11,267,500	1,135,000	1,243,210	547,340	200,000		677,847	333,108	214,232	344,739

Cost in excess of billings $200,000
Earned income unbilled 344,739
To balance sheet as asset $544,739

Imperturbable Constructors, Inc.

Balance Sheet

4/30/98

Current Assets			**Current Liabilities**		
Cash in bank	111,414		Equipment notes—1 yr	37,500	
Marketable securities	165,000		Bank notes–short-term	100,000	
CVLI	10,300		Accounts payable	74,900	
Prepaid taxes	4,000		Billings in excess of costs and estimated earnings	214,232	
Accounts receivable	164,300		Accrued income taxes	22,000	
Retainages receivable	44,000		Accrued misc. taxes	4,200	
Cost & estimated earnings in excess of billings	544,739		Accrued bonuses & salaries	5,300	
Inventory—WIP	13,800				
TOTAL CURRENT ASSETS		1,057,553	TOTAL CURRENT LIABILITIES		458,132
Fixed Assets			**Long-Term Liabilities**		
Accounts rec.—Employees	1,500		Equipment notes—Long-term	41,500	
Equipment (book value)	93,700		Subordinated officer's note	26,500	
Land	137,500		Deferred income taxes	10,500	
TOTAL FIXED ASSETS		232,700	TOTAL LIABILITIES		536,632
Other Assets			**Capital Stock**		
Unamortized			Common stock	150,000	
Leasehold improvements			Additional paid-in capital	100,000	
			Retained Earnings		503,621
Total Assets		1,290,253	Total Liabilities & Capital		1,290,253

169

Data Sheet
Imperturbable Constructors, Inc
4/30/98 FYE Statement

Capital stock (common)	$150,000	
Reserve—Doubtful accounts receivable	23,000	
Accrued miscellaneous taxes	4,200	
Georgia Power Bonds	165,000	
Equipment notes due in one year	37,500	
CSVLI	10,300	
Equipment (book value)	93,700	
Retainages receivable	44,000	
Accounts receivable	187,300	
Inventory—Work in progress	13,800	
Cash in bank	111,414	
Accrued bonuses and salaries	5,300	
Accounts payable	74,900	
Accrued income taxes	22,000	
Equipment notes—Due beyond one year	41,500	
Bank note payable (short-term)	100,000	
Deferred income taxes	10,000	(from depreciation)
Notes payable—Officers	26,500	(subordinated)
Prepaid income taxes	4,000	
Land	137,500	
Unamortized leasehold improvements	55,000	
Accounts receivable—Employees	1,500	
Additional paid-in capital	100,000	
Cost and estimated earnings over billings	544,739	
Billings over costs and estimated earnings	214,232	
Retained earnings	503,621	

PROBLEM 5

Good Ole Boy Construction Co.

The attached statement of Good Ole Boy Construction Company is presented to you on a completed contract basis just as their apprentice bookkeeper prepared it from the books. The same accounting method is used for both financial and tax reporting. From the information available, convert this statement to a percentage of completion basis, which would include an adequate tax reserve using a 36% tax rate factor on income earned. Reclassify anything as you feel necessary.

There are several important points to clear up before an accurate analysis can be made and this information is found on the last sheet of this problem—but jot down your own questions first before referring to it.

Also determine:

1. Current and debt-to-equity ratios on both accounting bases.
2. With $1,000,000 in uncompleted work, would you favorably entertain a bid bond request for a $3,500,000 bid going in next week to the Kingdom of Saudi Arabia for an underground parking garage in Abu Dhabi?
3. The caliber (quality and accuracy) of the financial presentation by the bookkeeper on a scale of 1 to 5.
4. Other underwriting information necessary.
5. Opinion of gross profits being realized.

Good Ole Boy Construction Co.
Work in Progress Schedule
1/31/98

Job Description	Owner	Contract Price Including Change Orders	Estimated Cost When Bid	Total Billed to Date Including Retainage	Total Costs to Date	Revised Estimated Costs to Complete
Bon Nuit Funeral Home	Alphsons Crematoria	8,429,614	7,981,741	8,429,604	7,880,714	10
Mia Pizza Parlor	Momma Mia	6,444,112	5,799,701	4,202,800	4,202,803	2,019,422
Totals		14,873,726	13,781,442	12,632,404	12,083,517	2,019,432

Complete Contracts Schedule
Completed Since Last FYE

Job Description	Owner	Final Contract Price	Total Cost	Gross Profit
Totals				

Good Ole Boy Construction Co.
Analysis of Work in Progress
1/31/98

1 Total Revised Costs	2 Original Estimated Gross Profit	3 Revised Gross Profit (Loss)	4 Billings in Excess of Costs	5 Costs in Excess of Billings	6 Ratio of Costs to Date to Total Costs	7 Income Earned	8 Billed Income Earned	9 Billings in Excess of Costs & Estimated Earnings	10 Earned Income Unbilled
Totals									

Good Ole Boy Construction Co.
Balance Sheet
1/31/98

Current Assets	Completed Contract	Percentage
Cash in bank	204,100	
West Virginia Turnpike bonds	5,760	
Due from affiliates	18,600	
Accounts receivable	146,700	
Earned estimates and retainage	606,000	
Costs in excess of billings	3	
Prepaid expenses	10,900	
TOTAL CURRENT ASSETS	992,063	
Fixed Assets		
Notes Receivable	41,800	
Prepaid federal income taxes	73,200	
Plant & equipment (net)	725,000	
Other Assets		
CSVLI	25,000	
Organization expense	6,500	
Total Assets	1,863,563	

Current Liabilities	Completed Contract	Percentage
Notes Payable	10,600	
Accounts Payable	410,600	
Billings in Excess of Costs	548,900	
Accrued Expenses	39,020	
Income Taxes Payable	120,800	
TOTAL CURRENT LIABILITIES	1,129,920	
Long-Term Liabilities		
Notes payable	45,270	
6¾% convertible subordinated debentures due 11/31/89	166,000	
TOTAL LIABILITIES	1,341,190	
Capital Stock		
Common stock	10,000	
Additional paid-in capital	199,400	
Retained Earnings	312,973	
Total Liabilities & Capital	1,863,563	
Working Capital	(137,857)	
Net Worth	522,373	

Questions That Should Have Been Asked

1. How do you check the latest trading price for West Virginia Turnpike bonds?
2. Has the 41,800 note receivable been collected since the statement date?
3. Should the 6,500 in organization expense be considered a tangible asset?
4. How are receivables turning over?

Answers

1. You can't. They have defaulted on payment of interest for the past 20 years. Eliminate entirely.
2. No. The maker is insolvent and filed for Chapter 13 bankruptcy five years ago. Eliminate entirely.
3. No. This is a deferred charge being written off against earnings. Eliminate entirely.
4. Great! Nothing over 60 days, except retainage. Allow as declared.

Answer to Problem 5

Good Ole Boy Construction CO.
Analysis of Work in Progress
1/31/98

1 Total Revised Costs	2 Original Estimated Gross Profit	3 Revised Gross Profit (Loss)	4 Billings in Excess of Costs	5 Costs in Excess of Billings	6 Ratio of Costs to Date to Total Costs	7 Income Earned	8 Billed Income Earned	9 Billings in Excess of Costs & Estimated Earnings	10 Earned Income Unbilled
7,880,724	447,873	548,890	548,890		100.0%	548,890	548,890		
6,222,225	644,411	221,887		3	67.5%	149,773	3		149,770
	10% gross profit	(3.4% gross profit)							
Totals									
14,102,949	1,092,284	770,777	548,890	3		698,663	548,893		149,770
		(5.18% gross profit)				× 36%			
							Deferred tax	Accrual	
						251,518			

176

Good Ole Boy Construction Co.
Balance Sheet
1/31/98

	Completed Contract	Percentage
Current Assets		
Cash in bank	204,100	204,100
West Virginia Turnpike bonds	5,760	
Due from affiliates	18,600	
Accounts receivable	146,700	146,700
Earned estimates and retainage	606,000	606,000
Costs in excess of billings	3	149,776
CSVLI		25,000
Prepaid federal income taxes		73,200
Prepaid expenses	10,900	
TOTAL CURRENT ASSETS	992,063	1,204,776
Fixed Assets		
Notes Receivable	41,800	
Prepaid federal income taxes	73,200	
Plant & equipment (net)	725,000	725,000
Prepaid expenses		10,900
Due from affiliate		18,600
Other Assets		
CSVLI	25,000	
Organization expense	6,500	
Total Assets	1,863,563	1,959,276

	Completed Contract	Percentage
Current Liabilities		
Notes payable	10,600	10,600
Accounts payable	410,600	410,600
Billings in excess of costs	548,900	
Accrued expenses	39,020	39,020
Income taxes payable	120,800	120,800
6¾% convertible subordinated debentures due 11/31/89		166,000
Deferred income taxes		251,518
TOTAL CURRENT LIABILITIES	1,129,920	998,538
Long-Term Liabilities		
Notes payable	45,270	45,270
6¾% convertible subordinated debentures due 11/31/89	166,000	
TOTAL LIABILITIES	1,341,190	1,043,808
Capital Stock		
Common stock	10,000	10,000
Additional paid-in capital	199,400	199,400
Retained Earnings	312,973	706,068
Total Liabilities & Capital	1,863,563	1,959,276
Working Capital	(137,857)	206,238
Net Worth	522,373	915,468

Other determinations:

1. Basis	Current Ratio	Debt-to-Equity Ratio
Completed contract	1:1.39	2.56:1
Percentage of Completion	2.20:1	1.14:1

2. With working capital on a percentage basis of $206,238 and only a 1.20:1 current ratio, there would not be enough to support a $4,500,000 work program (4.5% working capital case). If the holders of the $166,000 subordinated debentures agreed to extend the maturity for another year or two, the $372,238 working capital that would result from such extension would provide a 8.27% working capital case. Even without the hazards associated with working in a foreign country, this would make an unattractive surety risk.

3. 1 on the scale. The bookkeeper showed the debentures that would mature three months after the statement date as a long-term liability. Prepaid federal taxes should have been carried as a current asset. The note receivable in litigation should have been noted to that effect. The worthless West Virginia Turnpike bonds should also have been eliminated and the statement noted to this effect.

4. If the financial condition did warrant further consideration of the proposed bid, we would want full job details including:
   - Description of the work
   - Phases to be subcontracted and subbonding requirements to be imposed
   - Completion time
   - Penalties or liquidated damages

5. Combined gross profit percentage of 5.18% would be considered marginal and the substantial profit erosion on the Pizza Parlor at only 67.5% completion would certainly be a negative factor.

PROBLEM 6

Unscrupulatio Developers, Inc.

On 6/25/98 this firm has been requested by its bank to have its CPA make a conversion of the attached 9/30/97 modified cash basis statement to percentage of completion. Because time was of the essence and the CPA wasn't available, the contractor furnished the commercial loan officer (you) with the following information as of the statement date and told you to "have at it."

Accounts receivable	$4,860,500
Earned estimates/retainage	3,444,100
Accounts payable	2,421,300
Due subcontractors	3,005,888

The attached work in progress schedule was also furnished.

Although you were being asked to entertain a sizable unsecured revolving line of credit and couldn't do so on the basis of this "patchwork" financial exercise, you could at least make some rough determinations as to whether such a loan was feasible. Before beginning, however, there were questions that had to be answered in order to arrive at something resembling the true financial condition of your applicant. Your conversion would follow the customary classification of assets and liabilities for analytical purposes and, therefore, not necessarily in accordance with generally accepted accounting principles. Jot your questions down before referring to the question and answer sheet found at the end of this problem.

Once you have made your conversion, you will also want to take a look at the ratios and enter these on the lines provided just under the balance sheet, as they apply only to the percentage statement. Don't forget the timing difference in recognition of earnings in the two accounting methods when making a provision for a deferred tax liability. For the sake of this exercise assume deferred taxes are $1,199,445.

The urgency of this loan arises from an excellent opportunity to bid on a new synagogue in Tehran, Iran with an estimate of $25,000,000 in U.S. dollars. If awarded the contract, Unscrupulatio would have substantial up-front mobilization costs and the company didn't want to bid the job if it wasn't sure that the loan would be approved.

With your questions answered, the statement conversion made, and the analysis completed, what are your conclusions and recommendations to the loan committee?

Unscrupulatio Developers, Inc.
Work in Progress Schedule
9/30/97

Job Description	Owner	Contract Price Including Change Orders	Estimated Cost When Bid	Total Billed to Date Including Retainage	Total Costs to Date	Revised Estimated Costs to Complete
Romano Towers	Mario Speculato	10,748,010	9,773,209	7,566,015	7,669,855	3,708,404
Serenity Apartment	Serenity Inc.	12,994,717	11,695,246	10,889,381	10,073,981	815,666
Totals		23,742,727	21,468,455	18,455,396	17,743,836	4,524,070

Completed Contracts Schedule
Completed Since Last FYE

Job Description	Owner	Final Contract Price	Total Cost	Gross Profit
Totals				

Unscrupulatio Developers, Inc.
Analysis of Work in Progress
9/30/97

1 Total Revised Costs	2 Original Estimated Gross Profit	3 Revised Gross Profit (Loss)	4 Billings in Excess of Costs	5 Costs in Excess of Billings	6 Ratio of Costs to Date to Total Costs	7 Income Earned	8 Billed Income Earned	9 Billings in Excess of Costs & Estimated Earnings	10 Earned Income Unbilled
Totals									

Unscrupulatio Developers, Inc.
Balance Sheet
9/30/97

Current Assets	Modified Cash	Percentage
Cash in bank	121,000	
Marketable securities	164,800	
Inventory	650,000	
TOTAL CURRENT ASSETS	935,800	
Fixed Assets		
Plant & equipment	126,000	
Other Assets		
CSVLI	45,000	
Goodwill	25,000	
Total Assets	1,131,800	

Current Liabilities	Modified Cash	Percentage
Current equipment notes	50,000	
Income taxes due	4,575	
Accruals		
TOTAL CURRENT LIABILITIES	54,575	
Long-Term Liabilities		
Equipment notes	25,500	
TOTAL LIABILITIES	80,075	
Capital Stock		
Common stock	150,000	
Retained Earnings	901,725	
Total Liabilities & Capital	1,131,800	
Current ratio _____		
Debt-to-equity ratio _____		
(Only for % analysis)		

Questions/Answers

1. What do marketable securities consist of?

Black and Decker stock, which was being traded on the NYSE at just about the same price as that of the statement date.

2. Have any of the notes receivable been collected since the statement date?

All have.

3. What does inventory consist of and how is its value declared?

$125,000 in general warehouse materials accumulated over the past three years and $525,000 purchased for the Romano Towers job. The $125,000 is valued at the lower of cost or market (allow 50% as current asset) and the $525,000 is at cost.

4. What has been the largest job completed by Unscrupulatio to date?

Serenity Apartment for $13,000,000.

5. What is their uncompleted work program now?

$26,000,000.

6. With the word "Developer" in their corporate name, does this imply that there is some ownership interest by Unscrupulatio in their jobs in progress and past jobs?

Not exactly—Angelo Unscrupulatio is the sole stockholder in Serenity, Inc. and arranged for interim and permanent financing for the full development costs, including the land. There is a covenant in the construction loan agreement that requires preleasing of at least 75% of the units before the lender will fund more than 50% of the amount committed. Not only has Mr. Unscrupulatio personally guaranteed this loan, but the lender has taken the corporate guarantee of Unscrupulatio Developers, Inc. At the point when 50% of the loan had been disbursed, only 30% of the units had been preleased and funding of the remaining loan agreement was withheld. Serenity, Inc., therefore, couldn't pay Unscrupulatio Developers for all of the work they had performed—hence the very sizable unbilled costs and earnings on the job. They may never be recovered, particularly should the lender foreclose.

7. What was the full construction loan commitment on Serenity, how much had been drawn down against it, what additional costs had been incurred since 9/30/97, and how did all of this impact the contractor's financial condition?

The total commitment was $12,500,000. $6,250,000 had been drawn and full costs of $11,378,259 had been incurred. Mr. Unscrupulatio had been able to personally raise just over $2,000,000 and fully exhausted Unscrupulatio Developers, Inc.'s corporate working capital. The Serenity House was completed and the lender did convert the construction loan balance to 30-year permanent financing—but only when the occupancy rate reached 70% after completion.

Unscrupulatio Developers, Inc.
Balance Sheet
9/30/97

Current Assets	Modified Cash	Percentage
Cash in bank	121,000	121,000
Marketable securities	164,800	164,800
Notes receivable		4,860,500
Accounts receivable		3,444,100
Earned estimates and retainage		
Costs & estimated earnings in excess of billings		1,131,790
CSVIL		45,000
Inventory	650,000	587,500
Prepaid insurance		
TOTAL CURRENT ASSETS	935,800	10,354,690
Fixed Assets		
Plant & equipment	126,000	126,000
General inventory (50%)		62,500
Other Assets		
CSVLI	45,000	
Goodwill	25,000	Eliminate
Total Assets	1,131,800	10,543,190

Current Liabilities	Modified Cash	Percentage
Current equipment notes	50,000	50,000
Accounts payable		2,421,300
Due subcontractors		3,005,888
Billings in excess of costs & estimated earnings		526,409
Deferred income taxes		1,199,445
Income taxes due	4,575	4,575
Accruals		
TOTAL CURRENT LIABILITIES	54,575	7,207,617
Long-Term Liabilities		
Equipment notes	25,500	25,500
TOTAL LIABILITIES	80,075	7,233,117
Capital Stock		
Common Stock	150,000	150,000
Retained Earnings	901,725	3,160,073
Total Liabilities & Capital	1,131,800	10,543,190

Current ratio 1.44:1
Debt-to-equity ratio 2.22:1
(Only for % analysis)

Unscrupulatio Developers, Inc.
Analysis of Work in Progress
9/30/97

	1 Total Revised Costs	2 Original Estimated Gross Profit	3 Revised Gross Profit (Loss)	4 Billings in Excess of Costs	5 Costs in Excess of Billings	6 Ratio of Costs to Date to Total Costs	7 Income Earned	8 Billed Income Earned	9 Billings in Excess of Costs & Estimated Earnings	10 Earned Income Unbilled
	11,378,259	974,801	(630,249)		103,840	67.4%	(630,249)[a]		526,409	
	10,889,647	1,299,471	2,105,070	815,400		92.5%	1,947,190	815,400		1,131,790
Totals	22,267,906	2,274,272	1,474,821	815,400	103,840		1,316,941	815,400	526,409	1,131,790

[a]Full impact of loss will be reflected on balance sheet as follows:

Costs in excess of billings for which no asset value will be given $103,840

Costs yet to be expended in excess of contract price available (Billings/Costs)   526,409, i.e.,

$$\frac{526,409}{630,249}$$

(This part of loss already absorbed)

Total contract price $10,748,010

Total billed to date   7,566,015

   3,181,995

Costs to complete   3,708,404

   (526,409)

1. Whatever form "spec" building takes, it leaves a lot to chance, particularly where commitments from financially sound "core" tenants are not arranged well in advance. Most lenders will impose some preleasing requirements before approving a loan for a private commercial building. The market dynamics in real estate investment often follow erratic and unpredictable "boom or bust" cycles. Through the mid 1970s, REITS (Real Estate Investment Trusts) were a very popular, highly leveraged vehicle for investing enormous sums in commercial real estate, with many prominent commercial banks overextending themselves in financing these ventures. When the "bubble burst," many lenders found themselves with staggering loan defaults and foreclosures.

2. While it is essential to thoroughly review a contract between contractor and owner, it is also highly advisable to know what other commitments or contingencies exist that could, at a later date, impose financial hardships on your client.

3. We had the act of "omission" with Mr. Unscrupulatio. Knowing full well the disastrous consequences of the Serenity underfunding debacle, he had chosen to let you find out for yourself. You, therefore, have a serious credibility problem with this individual, which ties in directly to the first of the three C's—Character.

Looking back at the statement conversion exercise, the following calculation may be of some help in following how the cash statement reconciles with the one on a percentage basis:

RECONCILEMENT BETWEEN CASH AND PERCENTAGE METHODS

Retained earnings—Cash	$ 901,725
Add	
Cost & earnings/billings	1,131,790
Accounts receivable	4,860,500
Earned estimates	3,444,100
Total	$10,338,115
Subtract	
Accounts payable	(2,421,300)
Due subs	(3,005,888)
Billings costs/earnings	(526,409)
Deferred taxes	(1,199,445)
Goodwill	(25,000)
Retained earning %	$ 3,160,073

PROBLEM 7

Enviable Constructors, Inc.

You are presented with the attached compiled financial statement (Column 1 on the Comparative Financial Analysis) of Enviable Constructors, Inc., prepared on a modified cash basis for both financial and tax reporting, along with other supporting documentation, and asked to consider a $15,000,000 bid for construction of a new courthouse in New Orleans, Louisiana.

Although this general contractor has never been a seeker of surety credit, they have successfully completed larger, more complex unbonded private contracts in Georgia and Alabama and received many awards for excellence from various owners and architects in the past.

Their trade reports reflects that this is a family-owned, third-generation firm dating back to 1916, trade payments to subs and suppliers are discount and prompt, and, because of excellent cash flow and financial management, there has never been an occasion where short-term bank credit has been needed or applied for.

The found is 89 years old and still active in conducting daily affairs. While his son took little interest in the business, he did try his hand for a short period as a project manager for Enviable, only to cost the firm hundreds of thousands of dollars in poorly managed jobs. The grandson is very keen on the construction business, just graduated from Auburn with a degree in building construction, and is his grandfather's only hope of having the business continue under family control. There is no personal indemnity available to the surety since the owners feel the corporation is financially sound enough to stand on its own.

The company is fortunate in having Pete Smith, a thoroughly experienced field supervisor, who has worked for Enviable for 50 years. Harold Jones, another loyal and devoted employee, had just passed away after 55 years of service. To assure continuity, the founder has dismissed his son and changed his will so that the grandson will inherit all of the company stock upon his death.

Upon reviewing the 2/28/98 fiscal year financial statements prepared by the company's bookkeeper, you immediately recognize that a cash statement is totally unacceptable, as it omits all accounts receivable and payable and recognizes income on strictly a cash-in/cash-out basis. The bookkeeper, Miss Judge, is another "old timer" with Enviable and has always been very diligent in manually posting receipts and disbursements to the single entry ledgers each week. Never in 30 years has she failed to reconcile their cash ledgers with the bank statement. As an alert, but novice, underwriter, you inform the contractor that: (1) you cannot accept an internally prepared fiscal statement, (2) cash statements are out, and (3) you must have at least a reviewed financial statement.

A completed contract financial statement has been recommended by the recently engaged CPA and you agree that it should be adequate. When it is received, however, you notice that it doesn't include a schedule breaking down job-by-job costs and billings that would let you reconcile the under- and overbillings on the balance sheet. At your request, the CPA furnishes this information, you reclassify certain balance sheet items on the completed contract statement according to your company's underwriting practices, perform an analysis, and submit it to underwriter Joe Noe in your home office for his opinion. You inform him that Enviable has only $20,000,000 in other uncompleted work and that the $35,000,000 program, if Enviable is awarded the New Orleans contract, would be well in line with working capital and net worth, based on the completed contract analysis.

Does that please Joe Noe? *No!* A five-page letter ensues, admonishing you to make a percentage of completion conversion of the completed contract statement by use of a *Work in Progress Schedule.* How else, Joe inquires, can a true evaluation of profit trends on these uncompleted jobs be accomplished? What made you think that the costs in excess of billings, plugged into the completed contract financial statement didn't actually represent losing jobs, where there might not be a recovery of these unbilled costs? Even with a good idea of what G&A expenses will be, you can't determine that the projected gross profit margins are adequate to cover the fixed expenses and provide a profit without using a percentage of completion analysis.

Chastened, though undaunted, you require a WIP Schedule and set about determining the estimated profit on each uncompleted job, how these compare with the original bid profit, what percentage of the job has been completed, and how that percentage translates into earned profit at the statement date in each case. With knowledge of the costs incurred and the *earned* profit, you can now determine which jobs have been billed in excess of their respective costs and earnings, and vice versa. As these totals are inserted on the balance sheet as current assets and liabilities and the completed contract method retained earnings are changed, you rationalize that there will possibly be greater revenues on the P&L (assuming overall profitability), and thus a corresponding increase in net profit before taxes. You also decide that a look at the deferred income for tax purposes would be in order, so that an accrual can be made, if necessary.

You have completed your percentage of completion analysis, as well as completed the current and debt-to-equity ratios. With the healthy cash flow Enviable has enjoyed for the past year, and their solid earnings record over many years, you are comfortable in strongly recommending home office authorization. What else could Joe Noe want?

Problem

1. Complete the work in progress analysis sheet from the work in progress schedule.
2. On the percentage of completion P&L, fill in the blanks by use of the work in progress schedule and analysis. You will have to assume that all costs and billings arose during the present accounting period.
3. From the completed P&L and WIP analysis, complete the last two columns on the balance sheet and the blanks on page 223. Be alert to whatever deferred income tax accruals are needed. If any, use a *35%* tax rate factor on untaxed income.
4. What was Joe Noe's decision? Why?

Enviable Constructors, Inc.
Work in Progress Schedule
2/28/98

Job Description	Owner	Contract Price Including Change Orders	Estimated Cost When Bid	Total Billed to Date Including Retainage	Total Costs to Date	Revised Estimated Costs to Complete
Convention center	City of Albany, GA	3,648,948	3,284,053	1,788,990	1,412,333	1,655,221
Municipal boiler plant	Oxford, AL	14,889,406	14,144,935	13,706,118	12,944,001	918,035
Office tower	Corp. of Engineers	21,693,032	21,042,241	18,774,009	18,906,046	2,353,125
Totals		40,231,386	38,471,229	34,269,117	33,262,380	4,926,381

Completed Contracts Schedule
Completed Since Last FYE

Job Description	Owner	Final Contract Price	Total Cost	Gross Profit
Totals				

Enviable Constructors, Inc.
Analysis of Work in Progress
2/28/98

1 Total Revised Costs	2 Original Estimated Gross Profit	3 Revised Gross Profit (Loss)	4 Billings in Excess of Costs	5 Costs in Excess of Billings	6 Ratio of Costs to Date to Total Costs	7 Income Earned	8 Billed Income Earned	9 Billings in Excess of Costs & Estimated Earnings	10 Earned Income Unbilled
Totals									

191

P&L Statement for Each Accounting Method as of 2/28/98

	Modified Cash	Completed Contracts	Percentage
Revenues			
Completed contracts	$65,967,603	$41,688,040	$41,688,040
Uncompleted contracts			
Total	65,967,603	41,688,040	
Cost of Revenues			
Completed contracts		39,603,638	39,603,638
Uncompleted contracts			
Total gross profit	2,813,120	2,084,402	
G&A Expenses	(2,679,734)	(2,679,734)	(2,679,734)
	133,386	(595,332)	
Other Income			
Gain on sale of fixed assets	174,802		
Interest income	94,506	269,308	269,308
Other Expenses			
Interest expense	(57,992)	(57,992)	(57,992)
Net Income Before Taxes	344,702	(384,016)	
Federal and State Taxes	(120,645)	(120,645)	(120,645)
Deferred Taxes			
Net Income After Taxes	224,057	(504,661)	
Retained Earnings 2/28/97	8,507,704	7,878,253	7,878,253
Retained Earnings 2/28/98	8,731,761	7,373,592	

Enviable Constructors, Inc.
Balance Sheet
2/28/98

Current Assets	Modified Cash[a]	Completed Contract	Percentage of Completion
Cash—Bank	2,488,167	2,488,167	
Marketable securities	1,684,665	1,684,665	
Accounts receivable		8,771,090	
Earned estimates & retainage		4,604,186	
Cost in excess of billings			
Cost estimated earnings in excess of billings			
Inventory at cost	66,409		
Prepaid insurance	476,114		
TOTAL CURRENT ASSETS	4,715,355		
Fixed Assets			
Plant & equipment (net)	6,376,748		
Prepaid income taxes	785,000		
TOTAL FIXED ASSETS	7,161,748		
Other Assets			
Leasehold improvements	316,020		
Total Assets	12,193,123		

Current Liabilities	Modified Cash	Completed Contracts	Percentage of Completion
Notes payable	520,000	520,000	
Current portion—Long-Term Debt	306,114	306,114	
Accounts payable		3,061,798	
Due subcontractors		10,664,910	
Billings in excess of cost			
Billings in excess of cost & estimated earnings			
Income tax due	120,645		
Deferred income tax			
Accrued FICA taxes	98,606		
TOTAL CURRENT LIABILITIES	1,045,365		
Long-Term Liabilities			
Notes payable—Long-term	948,653		
TOTAL LIABILITIES	1,994,018		
Stockholder's Equity			
Capital Stock			
Common stock	1,017,344		
Preferred stock	450,000		
Retained Earnings	8,731,761		
TOTAL LIABILITIES & EQUITY	12,193,123		

[a]Illustration purposes only—Average annual revenues cannot exceed $5,000,000 for cash basis tax reporting.

	Modified Cash	Completed Contract	Percentage of Completion
Current assets	_____	_____	_____
Current liabilities	_____	_____	_____
Working capital	_____	_____	_____
Net worth	_____	_____	_____
Current ratio	_____	_____	_____
Net quick ratio	_____	_____	_____
Debt-to-equity ratio	_____	_____	_____

Answers to Problem 7

Joe Noe Declined.   Because of the age of Enviable's founder, you would have to assume that he is nearing the period where his ability to conduct affairs will diminish. The grandson, although well educated, is just too green to effectively contend with the host of complications that arise all too frequently in running a construction firm. He might be able to lean on the experience of Pete Smith in the short term, but Pete's no spring chicken either, and after 50 years of service, he may justifiably feel he deserves to "hang up his pistols" and enjoy life.

Altogether, regardless of the corporate financial strength, excellent reputation, and spotless performance record, there isn't enough in place to assure successful continuity of operations—particularly where you might find the company carrying $30,000,000 in uncompleted work with a novice at the helm. This one aspect alone would be a reasonable basis for the declination. During the underwriting process, you might also have inquired about what life insurance was in force on the life of the founder, and if the corporation was named as beneficiary. In any event, money is not the central issue, and the founder probably didn't carry any "key man" life insurance anyway, because of his age.

Knowing that Enviable had historically conducted its operations in Georgia and Alabama, you should have had serious concerns about its venturing so far afield into a totally unfamiliar area, and you should have investigated the local conditions that could impede its progress (or ruin it). A quick call to your New Orleans office (if you have one), or to someone else with experience in working with contractors in that region, would probably have resulted in a warning from your contact to stay away unless Enviable had strong local support and knew the *right* people in that area. Labor unions, regional economy, and the subcontractor market (to mention only a few) should have been prime considerations before you recommended approval. This is not to say that a contractor should not test new waters, but this should be done only after the company fully understands the local "climate" and, with all of the facts before it, can convince you that it is a sound move.

While adequate cash flow has thus far negated the need for short-term borrowing from Enviable's bank, you never know when a large receivable could be hung up over a dispute with an owner, or other impediments to cash flow arise. Enviable should have established a working capital credit line to supplement short-term capital needs resulting from unforeseen contingencies—if only as something of a safety net.

You, also, might very well question the CPA's professional expertise in the field of construction accounting, as well as in other general accounting areas. He or she should have recommended the percentage of completion method—or at least furnished an uncompleted job schedule with the completed contract financial statement to begin with. Why were prepaid taxes classified as a noncurrent asset? Why did Enviable have to pay taxes? There was adequate cash available to pay down trade obligations for the purpose of creating negative cash flow on the tax return.

This is the computer age, where manual posting of bookkeeping entries, particularly for a firm of this size, is following the trail of the dinosaur. With only single entry ledger posting, you could probably assume that there is no specific job cost allocation, thus rendering a completed contract or percentage statement a virtual impossibility. If your assumption is correct, where did the 2/28/98 work in progress numbers come from? How reliable are they? Why didn't you discuss Enviable's internal systems in much greater depth, and satisfy yourself the firm had a good handle on costs, and were able to track line item expenses with budgeted job costs at least once a week? Why didn't the CPA set up the books for compatibility with a good computer system and strongly recommend purchasing one?

Personal indemnity of the founder, and possibly his wife, would certainly be a prime underwriting consideration, as well as that from the grandson, who stands to inherit all of the stock. Without it, this void presents even greater barriers to favorable consideration.

Moral of this Sad Episode: As you are aware from earlier discussions, successful underwriting of the contractor risk extends well beyond meeting the basic 3C test. In Enviable's case, we had adequate capital, sound character and, apart from the continuity problem, the capacity to undertake a $15,000,000 contract. Without these three bedrock elements to build an underwriting case upon, there is no point going any further—but, beyond these, as we have seen, there are many other considerations.

With the highly complex and litigious environment surrounding the construction industry today, only the most seasoned contractor with strong leadership, good organizational balance, deep financial pockets, and highly sophisticated accounting systems can hope to survive and prosper. In many areas, there are too many contractors chasing too few jobs, thus paring profit margins to bare bone levels, and leaving no cushion for "contingencies."

Underwriter Joe Noe has done his job well, as this declination could have saved his company the agony of a huge loss.

Enviable Constructors, Inc.
Analysis of Work in Progress
2/28/98

1 Total Revised Costs	2 Original Estimated Gross Profit	3 Revised Gross Profit (Loss)	4 Billings in Excess of Costs	5 Costs in Excess of Billings	6 Ratio of Costs to Date to Total Costs	7 Income Earned	8 Billed Income Earned	9 Billings in Excess of Costs & Estimated Earnings	10 Earned Income Unbilled
3,067,554	364,895	581,394	376,657		46.0%	267,441	267,441	109,216	
13,862,036	744,471	1,027,370	762,117		93.3%	958,536	762,117		196,419
21,259,171	650,791	433,861		132,037	89.3%	387,437			387,437
Totals									
38,188,761	1,760,157	2,042,625	1,138,774	132,037		1,613,414	1,029,558	109,216	583,856

Enviable Constructors, Inc.
P&L Statement for Each Accounting Method as of 2/28/98

	Modified Cash	Completed Contracts	Percentage
Revenues			
Completed contracts	$65,967,603	$41,688,040	$41,688,040
Uncompleted contracts			34,875,794
Total	65,967,603	41,688,040	76,563,834
Cost of Revenues			
Completed contracts		39,603,638	39,603,638
Uncompleted contracts			33,262,380
Total gross profit	2,813,120	2,084,402	3,697,816
G&A Expenses	(2,679,734)	(2,679,734)	(2,679,734)
	133,386	(595,332)	1,018,082
Other Income			
Gain on sale of fixed assets	174,802		
Interest income	94,506		
	269,308	269,308	269,308
Other Expenses			
Interest expense	(57,992)	(57,992)	(57,992)
Net Income Before Taxes	344,702	(384,016)	1,229,398
Federal and State Taxes	(120,645)	(120,645)	(120,645)
Deferred Taxes			(309,644)
Net Income After Taxes	224,057	(504,661)	799,109
Retained Earnings 2/28/88	8,507,704	7,878,253	7,878,253
Retained Earnings 2/28/89	8,731,761	7,373,592	8,677,362
Deferred Tax Computations			
Net profit before tax	1,229,398		
Tax rate factor	× 35%		
Less:	430,289		
Cash basis tax	120,645		
Deferred tax	309,644		

Enviable Constructors, Inc.
Balance Sheet Treatment

Current Assets		Current Liabilities		
Costs in excess of billings (Column 5)	$132,037	Billings in excess of costs and estimated earnings (Column 9)		$109,216
Earned income unbilled (Column 10)	583,856	To reconcile with total billings in excess of costs of (Column 4)	$1,138,774	
Costs and estimated earnings in excess of billings	715,893	Column 9 as above plus billed income earned (Column 8)	109,216 1,029,558 1,138,774	(Unearned)

P&L TREATMENT (CONTRACTS IN PROGRESS)

Computing earned revenue

Billed to date		$34,269,117
Less:		
Billings in excess of costs		1,138,774
		33,130,343
Plus: Costs in excess of billings		132,037
Plus: Income earned		1,613,414
Earned revenues		34,875,794

Earned revenues (see opposite)	$34,875,794	
Cost of earned revenues	33,262,380	
Gross profit	1,613,414	

Enviable Constructors, Inc.
Balance Sheet
2/28/98

Assets

Current Assets	Modified Cash[a]	Completed Contract	Percentage of Completion
Cash—Bank	2,488,167	2,488,167	2,488,167
Marketable securities	1,684,665	1,684,665	1,684,665
Accounts receivable		8,771,090	8,771,090
Earned estimates & retainage		4,604,186	4,604,186
Cost in excess of billings		132,037	
Cost estimated earnings in excess of billings			715,893
Prepaid income tax		785,000	785,000
Inventory at cost	66,409	66,409	66,409
Prepaid insurance	476,114		
TOTAL CURRENT ASSETS	4,715,355	18,531,554	19,115,410
Fixed Assets			
Plant & equipment (net)	6,376,748	6,376,748	6,376,748
Prepaid insurance		476,114	476,114
Prepaid income taxes	785,000		
TOTAL FIXED ASSETS	7,161,748		
Other Assets			
Leasehold improvements	316,020	316,020	316,020
Total Assets	12,193,123	25,700,436	26,284,292

Liabilities and Equity

Current Liabilities	Modified Cash	Completed Contracts	Percentage of Completion
Notes payable	520,000	520,000	520,000
Current portion—Long-term Debt	306,114	306,114	306,114
Accounts payable		3,061,798	3,061,798
Due subcontractors		10,664,910	10,664,910
Billings in excess of cost		1,138,774	
Billings in excess of cost & estimated earnings			109,216
Income tax due	120,645	120,645	120,645
Deferred income tax			309,644
Accrued FICA taxes	98,606	98,606	98,606
TOTAL CURRENT LIABILITIES	1,045,365	15,910,847	15,190,933
Long-Term Liabilities			
Notes payable—Long-term	948,653	948,653	948,653
TOTAL LIABILITIES	1,994,018	16,859,500	16,139,586
Stockholder's Equity			
Capital Stock			
Common stock	1,017,344	1,017,344	1,017,344
Preferred stock	450,000	450,000	450,000
Retained Earnings	8,731,761	7,373,592	8,677,362
Total Liabilities & Equity	12,193,123	25,700,436	26,284,292

[a]Illustration purposes only—Average annual revenues cannot exceed $5,000,000 for cash basis tax reporting (See P&L and XYZ Construction Co., Inc. example).

	Modified Cash	Completed Contract	Percentage of Completion
Current assets	$ 4,715,355	$18,531,554	$19,115,410
Current liabilities	1,045,365	15,910,847	15,190,933
Working capital	3,669,990	2,620,707	3,924,477
Net worth	10,199,105	8,840,936	10,144,706
Current ratio	4.51:1	1.16:1	1.25:1
Net quick ratio[a]	3.99:1	1.11:1	1.21:1
Debt-to-equity ratio	5.11:1	1.90:1	1.59:1

[a]Eliminate inventory and prepaid expenses from current assets.

ADDITIONAL QUESTIONS—CHAPTER 9

Work in Progress Analysis and the Surety's Conclusions

1. What inference could be drawn from an internally prepared six-month interim statement, reflecting a 30% gross profit, when the preceding audited FYE statement by an independent CPA reflected only 15%, and historical gross margins had been in the 13–18% range?

_____

_____

2. Indicate True or False to the following assertions, and provide the correct answer to the false statements.

   A. To determine percentage of completion overbillings for the balance sheet, the cost incurred and unbilled earned profit are subtracted from billings to date on the WIP analysis.

   T _____ F _____

_____

_____

   B. For the balance sheet, costs and estimated earnings are a combination of costs incurred, earned income billed, and unearned income billed on the WIP analysis.

   T _____ F _____

_____

_____

   C. "Plowback" or "job borrow" are grounds for reducing working capital and net worth by the amount overbillings exceed the total estimated profit on a job.

   T _____ F _____

_____

_____

D. Billings in excess of costs is the basis upon which tax accruals are established under GAAP.

T _____ F _____

_____

_____

E. The cost-to-cost ratio approach to recognizing earned profits for percentage of completion accounting is the most equitable means of leveling out the peaks and valleys in tax liability when the Modified Cash method is used for tax reporting.

T _____ F _____

_____

_____

F. It is critically important to know the exact percentage of a completion for a job that will lose money so that the loss may be apportioned between one accounting period and another.

T _____ F _____

_____

_____

G. An accurately prepared WIP analysis by a CPA, reflecting increasing gross profit margins for all jobs in progress, as well as very healthy final profits on completed jobs, would enable an underwriter to be much more flexible in increasing a contractor's bonding line than for one showing only marginal or declining profits.

T _____ F _____

_____

_____

ANSWERS TO ADDITIONAL QUESTIONS—CHAPTER 9

1. In the absence of a substantial gross profit "windfall," the general presumption would be that profits were grossly overstated—pointing to inadequate cost records, overly optimistic and poorly founded estimates of remaining costs to complete and, in general, a very questionable accounting system. A knowledgeable contractor should have picked up the substantial disparity in earnings, and not submitted the statement to the surety until it was corrected, or until he or she could fully explain it.

2. A. False. The cost incurred and earned income would be subtracted from billings to date—or on the WIP analysis, billings in excess or costs less earned income.

   B. False. Costs incurred plus earned income billed and unbilled.

   C. False. They are already included as a current liability under billings in excess of costs and estimated earnings.

D. False. Income earned.

E. False. Percentage of completion profits have no bearing on cash basis tax reporting.

F. False. The percentage of completion is meaningless. The loss is fully recognized in the year discovered.

G. False. Few, if any, surety companies would increase a contractor's bonding line without a full statement—and then only on the basis of the contractor's FYE statement.

BIBLIOGRAPHY

CPCU 8 Course Guide, Accounting and Finance.

James Dan Edwards, Roger H. Hermanson, R. F. Solmonson, and Peter R. Kensicki. *Introduction to Accounting,* 2nd edition (Malvern, PA, American Institute, 1991).

ABOUT THE AUTHOR

With over 40 years experience as a practitioner in the fields of commercial bank and surety credit, Richard C. Lewis is one of the most respected names in the surety area of the insurance community. For 27 years he served in several top management positions for both Maryland Casualty Company and Fidelity and Deposit Company of Maryland in New York City, Washington, D.C., and Atlanta, Georgia.

Upon his official "retirement" in 1989, Richard focused his exemplary skills on surety consulting and writing. He has composed and edited material for the 1992–1993 academic CPCU program, in addition to sharing and demonstrating his expertise through this book. Most recently he has been requested to testify as an expert witness in a sizable contractor/surety case, currently in litigation.

In this book Richard has developed the broadest possible treatment of this complex credit discipline. He has integrated its legal, financial, construction, marketing, and banking components with a generous seasoning of contemporary underwriting philosophy. He has also endeavored to focus on realistic application of surety principles—thus enabling the reader to equate abstract theory with the pragmatic side of what is often a very sensitive and highly personal relationship between a contractor and his or her surety.

Index